AF594978

District Energy System Design

District Energy System Design

Simulation, Optimization and Decision Support

Special Issue Editors

Christian Inard

Jérôme Le Dréau

MDPI • Basel • Beijing • Wuhan • Barcelona • Belgrade • Manchester • Tokyo • Cluj • Tianjin

Special Issue Editors
Christian Inard
La Rochelle University
France

Jérôme Le Dréau
La Rochelle University
France

Editorial Office
MDPI
St. Alban-Anlage 66
4052 Basel, Switzerland

This is a reprint of articles from the Special Issue published online in the open access journal *Energies* (ISSN 1996-1073) (available at: https://www.mdpi.com/journal/energies/special_issues/DESD_SODS).

For citation purposes, cite each article independently as indicated on the article page online and as indicated below:

LastName, A.A.; LastName, B.B.; LastName, C.C. Article Title. *Journal Name* **Year**, *Article Number*, Page Range.

ISBN 978-3-03936-366-7 (Hbk)
ISBN 978-3-03936-367-4 (PDF)

Contents

About the Special Issue Editors

Christian Inard has been Professor at La Rochelle University (France) since 1998. He obtained his PhD with the thesis entitled "Thermal Coupling between Heating Systems and Dwellings" in 1988 and his Accreditation for Supervision of Research in 1996. His research activities are focused on design of low-energy buildings, indoor thermal comfort and air quality, HVAC systems modeling and control and urban microclimate modeling and experiments. He is author or co-author of more than 150 international papers and communications, and he has supervised 32 PhD candidates. Since 2011, he has been Dean of Faculty of Sciences and Deputy Head of the Institute for Research on Urban Sciences and Techniques.

Jérôme Le Dréau has been Assistant Professor at La Rochelle University (France) since 2015, before which he completed his PhD thesis and a 2-year postdoc at Aalborg University (Denmark). His research activities are related to energy flexibility of buildings, adaptive façades and low-temperature heating and cooling systems. He has published more than 25 articles through various journals and conferences. Jérôme Le Dréau has also participated in several international projects related to the energy flexibility of buildings, such as IEA EBC Annex 67 and Annex 82.

Preface to "District Energy System Design"

The integration of energy efficiency into the urban planning process is paramount in view of the current context of energy and environmental transition. However, reducing the energy footprint at the district level is a new approach that must be complemented by specific developments. Thus, the contributions related to the research of the best energy concepts cover three main topics: the simulation, the optimization procedure and finally the decision-making method. Towards this end, dynamic simulation tools based on physical models (occupant, building, production, networks, etc.) at the urban scale must be developed in order to represent the behavior of all energy flows of a district integrated in its environment. Then, specific characteristics of the problem complexity should be studied by means of a multi-objective and multi-stage optimization procedure. This cross-cutting approach could combine energy, economic and environmental aspects in order to create configurations that guarantee the best overall performance. Lastly, the selection of the preferential action could be achieved by the use of multi-criteria analysis methods that provide the planners with all the data that they need, enabling them to make the choice that best meets their expectations. This Special Issue includes five articles focused on these topics. The subjects covered are related to demand-side management, thermal flexibility, distribution network planning, multi-criteria decision-making, optimization of meshed distribution network, multi-carrier energy systems, combined optimal design, seasonal thermal energy storage, generation district heating networks, thermal energy storage and vehicle-to-grid for peak shaving, among others. All the papers in this Special Issue have been peer reviewed and subjected to the editorial standards of Energies. All our warmest and most sincere thanks go to all the authors for their timely response and careful revisions. They have produced very valuable contributions and provided high-quality materials for this Special Issue. Finally, we would like to sincerely thank the anonymous referees for their voluntary work and expert review.

Christian Inard, Jérôme Le Dréau
Special Issue Editors

Article

Integrated Optimal Design and Control of Fourth Generation District Heating Networks with Thermal Energy Storage

Bram van der Heijde [1,2,3], Annelies Vandermeulen [1,2,3], Robbe Salenbien [1,3] and Lieve Helsen [1,2,*]

1 EnergyVille, Thor Park 8310, 3600 Genk, Belgium
2 Department of Mechanical Engineering, KU Leuven, Celestijnenlaan 300, Box 2421, 3001 Leuven, Belgium
3 VITO NV, Boeretang 200, 5800 Mol, Belgium
* Correspondence: lieve.helsen@kuleuven.be

Received: 27 May 2019; Accepted: 10 July 2019; Published: 18 July 2019

Abstract: In the quest to increase the share of renewable and residual energy sources in our energy system, and to reduce its greenhouse gas emissions, district heating networks and seasonal thermal energy storage have the potential to play a key role. Different studies prove the techno-economic potential of these technologies but, due to the added complexity, it is challenging to design and control such systems. This paper describes an integrated optimal design and control algorithm, which is applied to the design of a district heating network with solar thermal collectors, seasonal thermal energy storage and excess heat injection. The focus is mostly on the choice of the size and location of these technologies and less on the network layout optimisation. The algorithm uses a two-layer program, namely with a design optimisation layer implemented as a genetic algorithm and an optimal control evaluation layer implemented using the Python optimal control problem toolbox called `modesto`. This optimisation strategy is applied to the fictional district energy system case of the city of Genk in Belgium. We show that this algorithm can find optimal designs with respect to multiple objective functions and that even in the cheaper, less renewable solutions, seasonal thermal energy storage systems are installed in large quantities.

Keywords: optimal design; optimal control; district heating; district energy systems; genetic algorithm; seasonal thermal energy storage; renewable energy

1. Introduction

Our energy system is one of the main contributors to the ever-increasing greenhouse gas (GHG) emissions, which calls for the identification of solutions within this sector. In particular, heating and cooling in residential and service buildings contribute to no less than 40% of the total final energy requirements in Europe [1]. Currently, 75% of the heating and cooling demand in buildings in the European Union (EU) (including industrial processes) is met by purely fossil resources [2], while of the remaining fraction 11% is provided by bio-mass, although these are usually polluting wood-stoves. Another 7% uses nuclear energy as a source through electricity and only the remaining 7% of heating and cooling come from 'truly' renewable sources, such as hydro, wind, solar and geothermal power. This makes the heating and cooling sector an ideal candidate to tackle the problem of both energy demand and GHG emissions in an efficient way. An often suggested solution is that of district heating and cooling systems to provide the heat and cold demand using centralised production with a large potential for reusing residual heat sources. Moreover, the most modern district heating and cooling systems (of the so-called *fourth generation*) allow for the inclusion of many low-temperature sources, thermal energy storage (TES) and prosumers that inject heat or cold surpluses back into the network [3].

Dahash et al. [4] provided a comprehensive overview of large-scale thermal energy storage systems, concluding that although not necessarily the most cost-effective, tank and pit storage systems are often the most practical to install. They also found a clear gap in the research towards system integration of these seasonal storage systems in terms of modelling both accurately and in reasonable time.

In particular, Paardekooper et al. [5] calculated that switching to a large share of district heating in the European energy system—including only established technologies—enables a reduction by 86% of the CO_2 emissions compared to the levels of 1990 but also that district heating can cost-effectively provide at least half of the heating demand in 2050 in the 14 countries that are studied in the Heat Roadmap Europe projects. This reduction is the result of a switch to renewable and residual energy sources (R^2ES), enabled by the heat transport provided by district heating. Moreover, according to Lund et al. [6], (seasonal) energy storage will be needed in a highly interconnected energy system, namely to bridge the fluctuations in the availability of renewable energy sources. They calculated that thermal energy and fuel storage are by far cheaper technologies than electrical (battery) storage.

Although it is important to know the potential of district heating and cooling systems, particularly in combination with TES systems, these previous studies only describe fourth generation district energy networks in general terms. Hence, to realise these innovative energy networks, three challenges need to be overcome. Firstly, the design of systems with large shares of fluctuating renewable energy sources will be much more complicated than that of present thermal networks. Secondly, identifying and implementing the right control strategy for a given network will be harder as well, due to fluctuating energy sources and large shares of energy storing components in the network, including energy flexibility. The third challenge follows from the first and second, namely the fact that the choice of control strategy will influence the optimal design and vice versa. This calls for an integrated strategy in which control and design are concurrently optimised.

1.1. Previous Studies on District Energy System Design

Söderman and Pettersson [7,8] made a topology optimisation algorithm for district energy systems (DES). The algorithm was based on a mixed integer linear program (MILP) for a district including thermal and an electric grid. Thermal energy storage was included in the optimisation, too. They limited the problem to eight representative time instances, namely typical daily and nightly operation conditions in the four seasons. Weber [9] integrated the optimisation of both design and control of poly-generation systems in DES with different energy carriers, but without considering TES. Again, the temporal detail remained limited. Weber used a bi-level solution strategy, where a master optimisation (evolutionary algorithm) chose the type, size and location of technologies to be installed in the network. The slave optimisation (mixed integer non-linear program) decides the layout of the network and the operational strategy.

Fazlollahi et al. [10] presented a multi-objective, non-linear optimisation strategy for DES including district heating and poly-generation, but without considering large-scale TES systems. They used a problem subdivision similar to that of Weber, where a master evolutionary algorithm varies the design parameters, and the proposed designs are evaluated by an MILP, which optimises the energy flows during 8 typical periods. Fazlollahi implemented an additional layer for the thermo-economic optimisation, and a post-processing step to assess the emissions of the proposed designs. The optimal results were summarised in Pareto-fronts according to system efficiency, total annual system cost and CO_2 emissions. In this study, the district heating supply and return temperatures were varied as a function of the ambient temperature.

Other studies combine the entire optimisation in a single mathematical problem, often an MILP. Dorfner and Hamacher [11] used this strategy to find the optimal lay-out and pipe size of district heating networks in Germany. This study omitted the operational aspect, instead only considering peak loads. Morvaj et al. [12] presented a single optimisation problem integrating design, operation and network layout for an urban energy system with 12 buildings. They considered one representative day for each month, averaging the electric and heat load profiles for a whole year. Falke et al. [13]

presented a similar multi-objective optimisation problem as Fazlollahi and Weber, but they considered a rule-based control flow for the operational layer, as opposed to an optimal control strategy.

In a wider energy system context, Patteeuw and Helsen [14] presented an integrated control and design optimisation algorithm for the design of the space heating and domestic hot water production system for residential buildings in the Belgian energy system, assuming a number of scenarios for the composition of the future electricity system. They used a single-layer MILP optimisation algorithm with representative weeks to reduce the temporal complexity of the optimisation problem. However, they found that this approach is very slow. They suggest that a scenario-based optimisation is more efficient than a full optimisation problem in which the scenario parameters are included as decision variables. This suggestion clearly points in the direction of a two-layered optimisation approach.

Lund and Mohammadi [15] presented a methodology to optimise the choice of insulation standard for pipes in thermal networks. Their method is split in two calculation tools: one to calculate different scenarios of heat loss behaviour in the thermal network, and the other where the energy flows in the larger system are optimised using EnergyPlan. An evolutionary design algorithm was coupled to EnergyPlan as the evaluation problem by Prina et al. [16]. Their focus was on the operation of regional energy systems to find both techno-economically feasible, as well as sustainable energy system designs. In a later step [17], they accounted for the long-term investment planning problem, considering the evolution of the price for different technologies and the remaining value of previously installed systems as they are replaced by more modern technologies.

Bornatico et al. [18] used a particle swarm optimisation (PSO) algorithm to optimise the thermal system of a Swiss single residential building (hence no DES or thermal network was considered). They aimed to optimise the size of a solar heating system, including a solar collector, storage tank and auxiliary power unit. In this study, the system was simulated in *Polysun*, coupled to MATLAB for the PSO implementation. Whether Polysun implements a heuristic or optimal control was not specified. The results of the PSO were compared to a genetic algorithm and the results were found to be similar. Ghaem Sigarchian et al. [19] optimised a hybrid microgrid including solar photovoltaic panels and concentrated solar power collectors, an organic Rankine cycle to convert heat to electricity, electric and thermal energy storage and a gas-fired backup generator. Both design variables (in the PSO) and a variable operation (in HOMER) were considered. The objective function was the energy tariff to be paid by the consumers in the network, which had to be minimised. The fitness evaluation function seems to be an optimal control problem implemented in HOMER, although this is not clearly explained.

In conclusion of the previous work, a clear pattern is that the optimisation algorithm is subdivided in two layers, where one layer is aimed at evaluating the operational aspect of a particular design—the lower layer or *slave algorithm*—and the other focusses on the exploration of the design parameter space—the upper layer or *master algorithm*. While there are subtle variations where for instance the slave algorithm also optimises part of the design variables, this general structure holds for most of the above discussed references. Still, a smaller number of studies use a completely integrated control and design algorithm, with a single layer that optimises both operational variables and design parameters. Clearly, this approach represents only a minority in the discussed studies and is only suited for design problems with a limited size and (temporal) complexity.

1.2. Novelty and Contribution

The aim of this paper is to develop an integrated design and control optimisation algorithm for future district heating systems with large shares of R^2ES and seasonal thermal energy storage. This algorithm is illustrated in a fictitious district heating system for an existing city in Belgium and the design results from the optimisation are studied in detail. Note that the focus is on the methodological contribution, rather than on the absolute numbers resulting from the case study.

Compared to pre-existing studies, this paper uses a two-layer approach, focussing on the integration of a higher-resolution full-year optimal control problem (OCP) as the lower-level optimisation layer, with particular attention paid to the high operation variability of future energy systems with distributed energy resources. In order to do so, a Python toolbox called `modesto` (see Vandermeulen et al. [20]) is used to set up these OCPs. We use an optimal selection of representative days compatible with seasonal thermal energy storage systems to reduce the calculation time. To our knowledge, this is also the first study in which a concurrent design of TES volumes, pipe diameters and heat generation systems is considered, together with a more detailed model of the district heating system.

2. Methodology

The optimisation framework is conceived as a two-layer integrated optimal design and control algorithm. In Section 2.1, the heat demand for space heating and domestic hot water is calculated deterministically and used as a fixed boundary condition for the algorithm. The slave optimisation is a linear optimisation which determines the optimal energy flows in the network for a given design, including the TES charging behaviour, implemented in `modesto`. This layer of the algorithm is explained in Section 2.2. The master optimisation is a genetic algorithm which looks for the optimal combination of design parameters, based on a number of objective functions. This layer is described in Section 2.3. Apart from the implementation, this section also summarises the available design choices for the chosen case study, as well as the considered scenarios.

2.1. Case Study

The optimisation algorithm is illustrated by means of a fictitious DES for the city of Genk in Belgium, called *GenkNet*. Spread over 9 neighbourhoods, 7775 building models were constructed based on geometric data for single family residential buildings. Although the network configuration and the choice of the connected neighbourhoods are hypothetical, the data with which the building models were constructed are real.

The building models are equivalent resistance-capacitance models based on the `TEASER FourElement` structure (see Remmen et al. [21]). Assumptions on the building materials and wall thicknesses were based on the building age, which was assumed fixed for all buildings belonging to one neighbourhood. The workflow to derive the building model parameters was developed by De Jaeger et al. [22]. The heat demand resulting from space heating and domestic hot water (DHW) production was simulated using a typical meteorological year for Belgium and stochastic occupant profiles as boundary conditions. The occupant profiles contain the space heating temperature set point and the DHW draw-off for every individual building and were derived using the `StROBe` toolbox (see Baetens and Saelens [23]). An ideal building heating system (neglecting the effects of the heating system temperatures on the heat injection in the building) was assumed. All buildings were simulated during a full year with a 900 s time step using a minimum energy objective (assuming a fixed cost for heat), after which the heat demand of all buildings belonging to one neighbourhood was summed and modelled as a single demand node in the network. The heat distribution network on the neighbourhood level is omitted, which means an underestimation of the total heat losses in the network. Instead, the neighbourhood is represented as a single node, connected to the backbone network through a single service pipe.

The resulting heat demand for the 7775 buildings amounts to 430.5 GWh per year, with an average energy use intensity of 210.8 kWh/m^2 per year. Cooling and electricity demand were not considered in this study.

The neighbourhoods were located alongside a central thermal network backbone, as indicated in Figure 1.

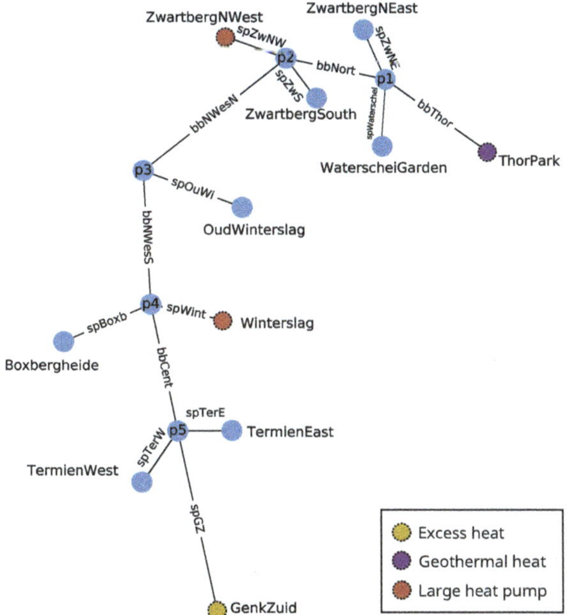

Figure 1. Network layout for the fictitious case study in Genk, Belgium. The heat demand of 9 neighbourhoods alongside a central backbone connection was aggregated per neighbourhood. Two additional nodes without heat demand, but with the option to install heat sources and/or thermal energy storage (TES) systems are situated at both ends of the backbone.

2.2. Operational Optimisation

We have chosen for a full-year OCP with a 2 h time-step for the evaluation of the DES performance with respect to a number of objective functions (see Section 2.3). The OCP optimises heat and mass flows in the network, the operation of the heat production systems and the charging behaviour of the TES systems in order to satisfy the heat demand of the neighbourhoods. This optimisation has a single objective function to minimise the operational cost. The choice for a minimal operation cost objective is based on the habit of operating real systems to maximise their profit. Except for a limited number of experimental systems, systems are seldom operated to minimise energy use or CO_2 emissions, unless this is linked to additional economic incentives. The choice for an optimal control strategy as opposed to a simulation based evaluation or a rule-based control strategy is justified by the quantification of the maximum potential of every design. This potential is always reached when we assume an optimal control strategy exists and can be implemented, whereas heuristic control strategies might penalise designs that are harder to control. As such, we arrive at a fair comparison of designs.

This OCP is implemented using the Python toolbox `modesto` (see Vandermeulen et al. [20]). This toolbox is built on top of Pyomo [24] and implements a library of linear optimisation models for common DES components and communicates with an optimisation solver to find the optimal operation strategy. More details about the used component models can be found in Appendix A. All models were either derived from literature or verified by the authors. The optimisation variables considered in the operational layer are the magnitude of heat and mass flows in all components, the thermal output of heating systems, possible curtailment of heat from the solar thermal collectors and the state of charge of the TES systems.

The solution time of the OCP is reduced by using representative days. Van der Heijde et al. [25] have developed a method to optimally select representative days and to restore the chronology such that the original data is approximated as closely as possible. This method is applied here. Based on a number of input time series, specified by the user, an optimal set of representative days is chosen, after which the algorithm determines for each day of the year which representative day it will be represented by. In this work, the chosen input time series were the aggregated heat demand for all neighbourhoods, the solar radiation on a unit surface area, the ambient temperature and the hourly electricity price. This method furthermore makes sure that seasonal effects in the TES systems are modelled accurately. We have limited the representative day selection to 12 days as this was shown to be sufficient to represent the full-year OCP with acceptable accuracy, see the conclusion made by van der Heijde et al. [25].

2.3. Design Optimisation

The design parameter values are varied in the upper layer. While we attempted to keep the slave optimisation linear, both to guarantee a global optimum and to limit the calculation time, non-linear effects do appear in the master optimisation problem. These non-linearities include: investment costs, which can vary with the size of the installed system; discrete decision variables, such as pipe diameters; and the calculation of the state-dependent heat loss from thermal storage tanks. This last phenomenon is caused by the use of a TES model in which the heat loss depends on the actual state of charge. On the other hand, this loss fraction also depends on the size of the TES system, which would render the optimisation problem bi-linear (see the derivation by Vandewalle and D'haeseleer [26]). To avoid this extra non-linearity, the design variable (namely the size of the TES systems) is treated as a constant parameter in the OCP and it is varied in the master optimisation.

We implemented the design optimisation algorithm as a genetic algorithm in Python using the DEAP (Distributed Evolutionary Algorithms in Python) toolbox [27]. The algorithm uses the NSGA-II (Non-dominated Sorting Genetic Algorithm II) selection operator [28]. Crossovers are handled by a simulated binary crossover operator, and for mutation, a polynomial bounded operator is used. Moreover, the genetic algorithm features a small probability of entirely reinitialising some parameters, which is a variation on the mutation operator. The Pareto-optimal solutions of all generations are stored in a *Hall of Fame*. Every new generation is initialised based on all non-dominated individuals, taken over all previous generations. As such, previous optimal solutions cannot be lost in the course of the evolution. A single optimisation run features 100 generations with 60 individuals. Each newly generated individual has a 95% mutation probability and a 70% crossover probability.

Every candidate design is evaluated as an instance of `modesto` with a minimal cost objective (see Section 2.2). Infeasible optimisation problems result in a high penalty objective value, such that these designs are not selected in the next generation. With `modesto`, the optimal control trajectory for all energy and mass flows in the network during one year is computed, such that the operational cost is minimal. The workflow of this genetic algorithm is illustrated in Figure 2.

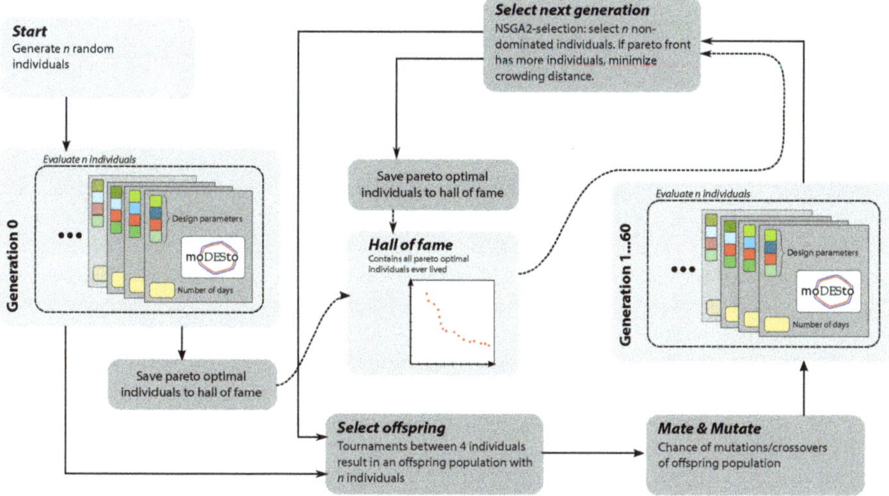

Figure 2. Flow chart illustrating the different steps in the genetic algorithm.

2.3.1. Objective Functions

The design was evaluated for optimality based on three objectives, namely the annual primary energy imported from outside the district energy system—used for the heat pumps, geothermal heating plant, possibly auxiliary heating needed for the production of DHW and network pumping power—, the total annualised costs and the CO_2 emissions. The total annualised cost c_a consists of the annual operation cost $c_{op,a}$ as calculated by `modesto`, increased by the annualised investment I_a and fixed annual maintenance cost $c_{maint,a}$:

$$c_a = I_a + c_{op,a} + c_{maint,a}. \tag{1}$$

The annualised investment cost I_a is calculated using the capital recovery factor:

$$I_a = I_{tot} \frac{(1+i)^\tau i}{(1+i)^\tau - 1}, \tag{2}$$

where I_{tot} stands for the total investment, and I_a for the annualised investment. τ denotes the economic lifetime of the technology and i is the interest rate, taken as 3%, assuming a public investment on a long term. This is in line with multiple studies, such as that of Möller and Werner [29], Nussbaumer and Thalmann [30] and Steinbach and Staniaszek [31]. The annualisation calculation using the capital recovery factor assumes that every component is replaced by an identical system at the end of its lifetime, and at the same investment. As such, variations in technology prices over time are neglected.

For maintenance ($c_{maint,a}$), the annual contribution is estimated to be a fixed fraction of the initial investment. This fraction, as well as the typical economic life time of various technologies, were derived from the EnergyPlan Cost Database [32]. All economic input data for the different technologies considered in this work is summarised in Appendix B.

In order to get a better grasp of the orders of magnitude of the objective functions—that is, annualised total costs, primary energy import and CO_2 emissions—we scale them with respect to the total annual heat demand of all neighbourhoods for space heating and DHW. This total heat demand amounts to 430.5 GWh per year. The resulting scaled variables are called the levelised cost of heat

(LCOH, expressed in EUR/kWh), the primary energy import share (PEIS, in %) and the CO_2 intensity (in kg CO_2/kWh).

2.3.2. Design Choices

The *GenkNet* case has a total of 9 neighbourhoods, 1 industrial node and 1 node with additional heat generation systems but no demand. The design exercise is left very open; the optimiser has to choose how many renewable resources for heat generation are installed at every node, where the TES systems are installed and how large they should be, and how much backup power is needed. The available design choices for the TES systems are listed in Table 1, the solar thermal collector (STC) arrays in Table 2, and the backup heat pumps and geothermal heating plant in Table 3. The maximum volume corresponds to the largest pit and tank TES systems currently found in literature. The maximum STC area corresponds to the available south-oriented roof area of the buildings in the neighbourhoods, however without accounting for previously installed systems, such as PV panels. At node *ThorPark*, a larger area is assumed to be available for the installation of an STC array.

Table 1. Available design choices for TES systems at the different nodes in GenkNet. All numbers are expressed in m^3.

Node	Component	Min	Max ($\times 10^3$)
Boxbergheide	PTES	0	200
OudWinterslag	PTES	0	200
TermienEast	PTES	0	200
ThorPark	TTES	0	12.5
WaterscheiGarden	PTES	0	200
Winterslag	PTES	0	200
ZwartbergNEast	PTES	0	200
ZwartbergNWest	PTES	0	200
ZwartbergSouth	PTES	0	200

Table 2. Available design choices for the installed STC array area per node. All nodes except *ThorPark* consider the total available South-oriented rooftop area of the considered buildings in that neighbourhood. All numbers are expressed in m^2.

Node	Min	Max ($\times 10^3$)
Boxbergheide	0	78.5
OudWinterslag	0	15.5
TermienEast	0	12.9
TermienWest	0	16.3
ThorPark	0	100.0
WaterscheiGarden	0	49.9
Winterslag	0	37.8
ZwartbergNEast	0	12.7
ZwartbergNWest	0	17.0
ZwartbergSouth	0	33.3

Table 3. Design choices for the nominal power of the central heat generation systems. The power is expressed in MW. The abbreviations "geo" and "hp" stand for geothermal heating plant and air source heat pump, respectively.

Node	Component	Min	Max
ThorPark	geo	0	40.0
Winterslag	hp	0	80.0
ZwartbergNWest	hp	0	80.0

In addition, the pipe diameters are also design decision variables. For the available diameters, the reader is referred to Tables A2 and A3. The smaller diameter pipes are implemented as twin pipes (up to DN 200), the larger pipe diameters as compound pipes. The investment costs for these pipes are discussed in Appendix B.3. The available diameters were derived from IsoPlus [33]. It is also an option to install no pipe at all at a specific network edge; this choice is represented by a 0 m diameter pipe.

All scenarios share the same network layout, as shown in Figure 1. Whereas the choice for the size of the heat pumps and the geothermal heating plant is handled by the optimisation algorithm, the availability of excess heat is fixed at 10 MW, constantly available throughout the year. Whether this resource is utilised or not is a question of the decision of the district heating connection between the node *GenkZuid* and the rest of the network, that is, is it worth the investment to make a connection to the industrial area from the city or not.

2.3.3. Scenarios

While most of the boundary conditions are fixed for the design algorithm, two of them are varied discretely and deterministically to establish their influence on the results, leading to a number of scenarios. The first boundary condition is the nominal temperature level in the network. Four options are available:

- 45–25 °C,
- **55–35 °C** *(base scenario)*,
- 65–45 °C and
- 75–35 °C.

Hence, most scenarios use a 20 K nominal ΔT with rather low supply temperatures, whereas the last scenario uses medium-high temperatures with a 40 K nominal ΔT.

The second scenario parameter is the cost of heat from the industrial excess heat source in the most southern node of the network. The heat prices considered are:

- 5 EUR/MWh,
- 10 EUR/MWh,
- **15 EUR/MWh** *(base scenario)* and
- 20 EUR/MWh.

These excess heat costs are substantially higher than the ones discussed by Doračić et al. [34], but they are chosen to be in line with the expected cost for an industrial company that needs to invest in a connection of its processes to a district heating system.

The different combinations of the two scenario parameters lead to a total of 16 optimisation runs to be performed. However, the focus will be on how the scenarios deviate from the reference scenario—55/35 °C with 15 EUR/MWh excess heat cost—leading to a total of 7 scenarios to be studied in detail.

3. Results

The emphasis of this paper is on the methodological contribution, namely the integrated design and control optimisation algorithm. The results presented in this section should mostly be interpreted as a proof of concept, rather than in absolute numbers.

3.1. Reference Case Results

As mentioned before, the case with a 55/35 °C temperature regime and an excess heat cost of 15 EUR/MWh is chosen as the reference case. This section shows a selection of visualisations of the optimal design results in order to make more sense out of the large amounts of output data. Firstly, we focus on the higher level, using only design parameters and yearly aggregated outcomes, but in a later stage we will also zoom in on the results on smaller time scales.

Genetic Algorithm Outcome

The solutions resulting from the genetic design algorithm, ranked by the objectives of LCOH and PEIS are shown in Figure 3. On the one hand, this figure shows the spread of all considered designs that turned out to be feasible in terms of their LCOH and PEIS values. The blue dots show the Pareto-optimal solutions, that is, the solutions that dominate the solution space. The three black markers respectively show the solutions with the lowest PEIS, the one with an approximately average PEIS measured over all non-dominated solutions and the one with the maximal PEIS. These three specific solutions will be treated in more detail in the next subsection. Note that the minimal and maximal PEIS solutions also represent the respective maximal and minimal LCOH solutions.

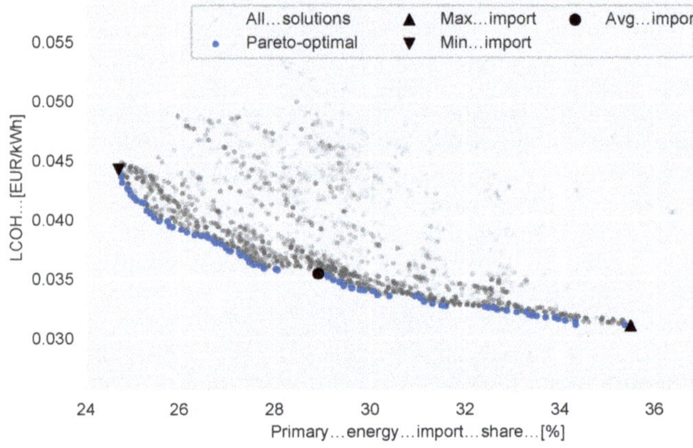

Figure 3. Scatter plot of all genetic design algorithm solutions for the reference design case.

Zooming in on the Pareto-optimal solutions only, we can plot the contributions of operational costs, investment and maintenance costs to the LCOH. This cost breakdown is presented in Figure 4, where we see that the contribution of the annualised investment takes up the largest share of the total annual costs of the system. As expected, as a larger amount of energy is imported from outside the network, the operational costs (representing in this case the cost of electricity use) increase linearly.

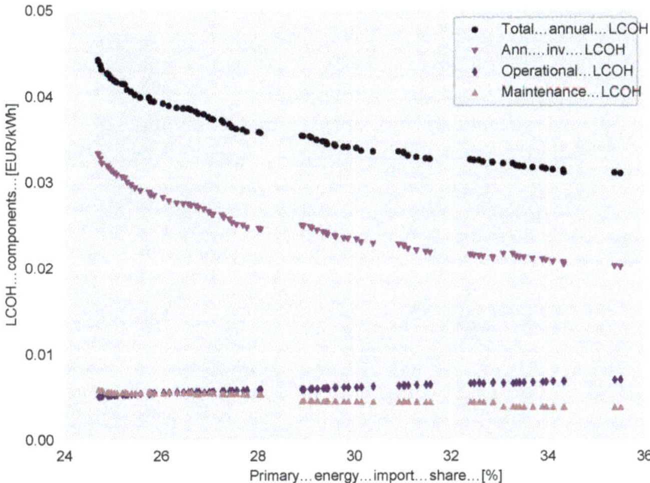

Figure 4. Different contributions to the total annual LCOH for the reference case, namely that of annual maintenance, operational cost and annualised investment cost. The annualised investment has the largest contribution.

The contribution of the different technologies is shown in Figure 5, showing indeed that the higher investment for the lower PEIS solutions is largely caused by a larger installed amount of TES and STC systems. The bar chart furthermore shows that the investment in auxiliary heating plants (i.e., the heat pumps and the geothermal plant) combined remains more or less constant. This does not mean that the installed auxiliary power remains constant as well, given the different prices per unit of thermal power for the two technologies. Finally, Figure 5 shows that the investment in the transport pipes remains more or less constant, too. On closer inspection (not shown here), we find that the lower PEIS solutions have marginally larger investments in the network, corresponding to wider pipes on average.

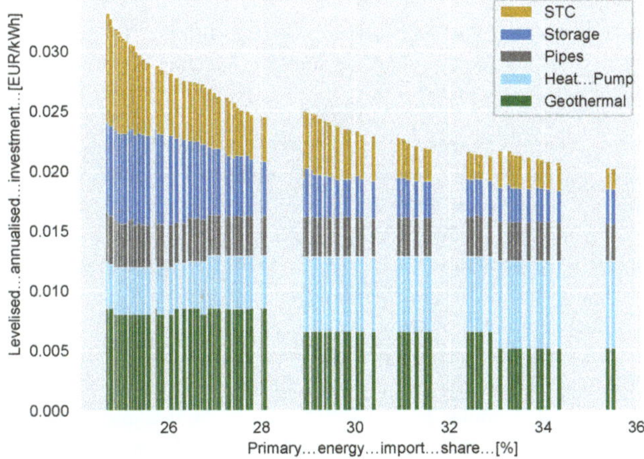

Figure 5. Bar chart showing the contribution of different technologies to the levelised annualised investment costs. The cost of the auxiliary domestic hot water (DHW) boilers has been omitted since it is the same for all solutions shown.

Figure 6 shows the distribution of the chosen sizes for the STC and TES systems in the network, split by network node and *energy range*. The energy range is a categorisation of the Pareto-optimal solutions based on their PEIS. The range of PEIS values for all Pareto-optimal solutions is split in three equal parts, denoting high, mid and low energy import. The plot shows three violin plots per neighbourhood and parameter, indicating the approximate distribution, as if the design parameters were distributed following some probability function. The plot gets wider where there is a denser distribution of observations for that parameter. The actual parameter values are indicated by the black dots inside the violin plots, which shows that the distribution is in fact very sparse, with a few dense spots here and there.

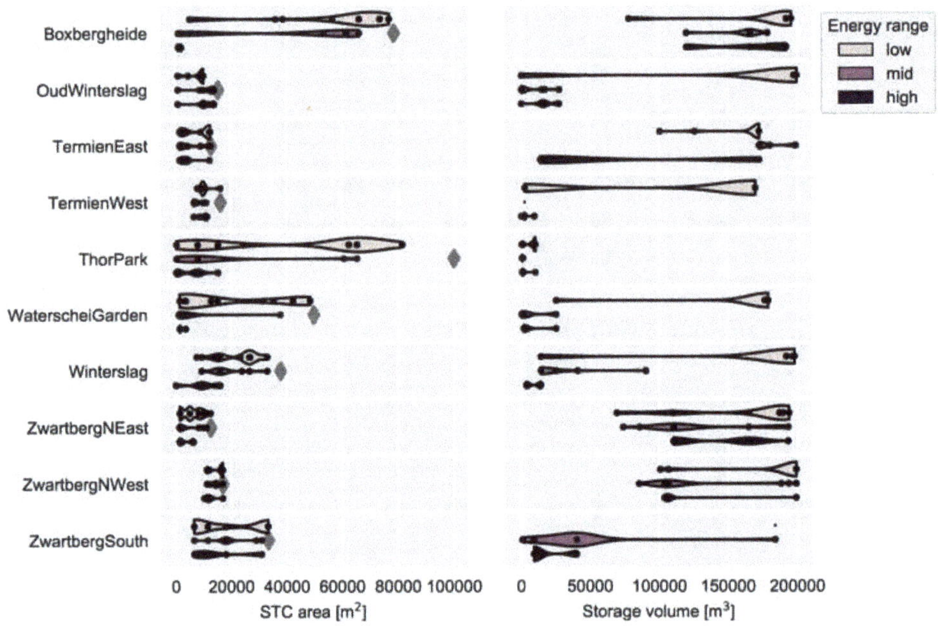

Figure 6. Distribution of solar and storage systems per network node. The Pareto-optimal solutions are split into high, mid and low energy import solutions. The left graph shows the maximum STC area per neighbourhood as the grey diamond.

The solutions with lower PEIS are expected to have a larger share of solar energy, which is only possible if enough STC area is installed, as well as TES volume. For example, *Boxbergheide*, *ThorPark*, *WaterscheiGarden* and *ZwartbergNEast* show this evolution of decreasing STC area with increasing imported energy. For the other neighbourhoods, the spread stays largely the same, although in most cases the average shifts to lower values. Only for *OudWinterslag*, no such shift can be distinguished. Another interesting observation is that more often than not, almost the maximum available area is exploited for the installation of STC systems.

Looking at the TES volumes, such a trend is not so easy to find. In the case of *OudWinterslag*, *TermienWest*, *WaterscheiGarden*, *Winterslag* and *ZwartbergSouth*, we can see that at least the low energy range has the largest storage volumes. However, if we compare this graph with the map (Figure 1), one interesting trend is that larger storage tanks tend to be installed as close as possible to locations where auxiliary heating plants (i.e., heat pumps and geothermal) are available. This is more obvious for the neighbourhoods of *ZwartbergNWest* and *Winterslag-Boxbergheide*. *ZwartbergNEast* seems to act as the storage hub for *ThorPark*, given the smaller distance between these nodes than the distance between

Waterschei Garden and *ThorPark*. The reason for the predominantly large volume at *ZwartbergNEast* can be explained by the limited storage potential at the *ThorPark* node.

When the distribution of the nominal power of the production systems is plotted per neighbourhood, we find that the variety in the solutions is rather low. Figure 7 shows that the size of the heat pump in *ZwartbergNWest* is positively correlated with the energy import range and the opposite is true for the geothermal heating plant. The heat pump at *Winterslag* has an almost constant nominal power.

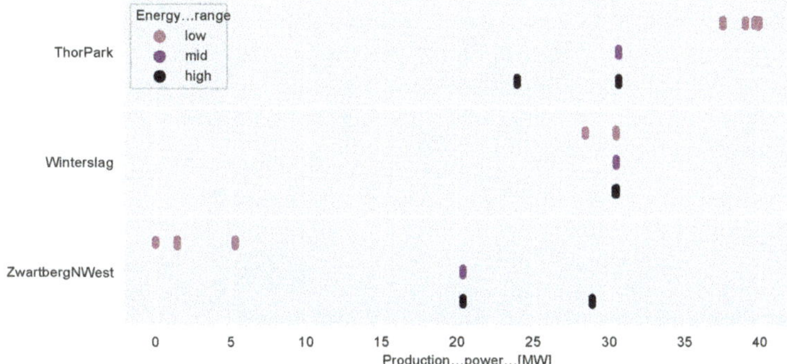

Figure 7. Spread of the nominal power of the chosen production systems, again subdivided according to the range of the energy import from outside the network. The heat pump at *Winterslag* is nearly always the same size, at *ZwartbergNWest* the size increases with the range of energy import and the geothermal heating plant decreases with increasing energy range.

A final plot of interest is the spread of the sum of all storage volumes compared to the total STC area in the network. This graph is shown in Figure 8, and a sigmoid distribution appears clearly. For lower STC areas, the storage volume appears to increase linearly with the STC area, until around $150 \times 10^3\,\mathrm{m}^2$, where the storage volume starts increasing very rapidly. As soon as the maximum storage volume is reached, the STC area can still increase, but mostly at a higher cost without much reduction in terms of import share.

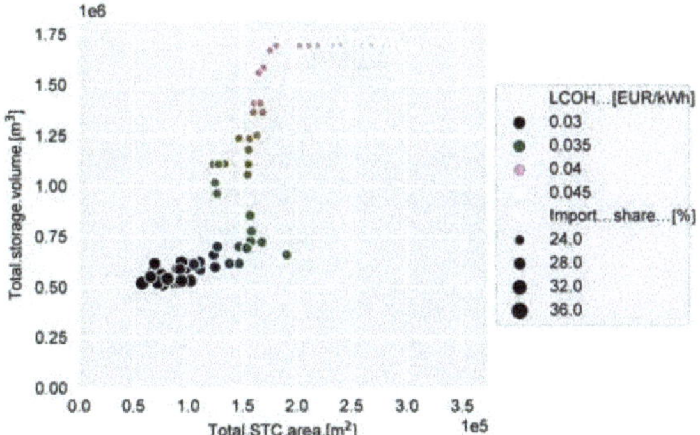

Figure 8. Distribution of total STC area compared to the total storage volume. The scatter points are coloured according to the LCOH, with darker values representing a lower LCOH than the lighter values. The larger the dot, the larger the import share (and thus the larger the grid dependence of the network). Note that the axes are limited by the minimum and maximum allowed design sizes.

3.2. Detailed Study of Highlighted Reference Solutions

In Figure 3, three particular solutions are indicated with a black marker. Two solutions respectively at the upper and lower end of the Pareto front were chosen, and one solution that is closest to the average PEIS of all Pareto-optimal solutions. For these designs, we investigate the heat sources in the network and the energy storage levels in more detail.

3.2.1. Maximum LCOH, Minimum PEIS

The first highlighted solution is the one with the lowest primary energy import from outside the network, but the highest LCOH. In the graph of the heat flow rates (the upper subplot in Figure 9), the black line indicates the part of the heat demand of the neighbourhoods which is supplied by the district heating network. We see that the excess heat production is used as a base load throughout most of the year, except for the Summer period, where most of the heat demand is met by a combination of solar thermal power and discharging of the TES systems.

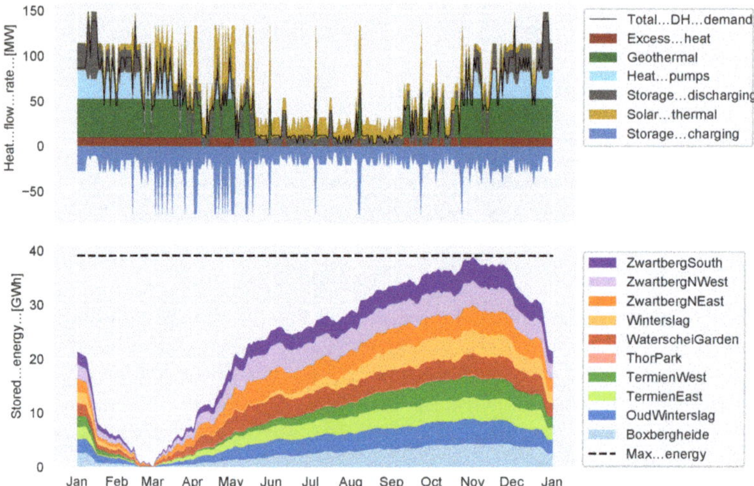

Figure 9. Full year time series for minimum PEIS design.

The geothermal plant is runner-up after the excess heat injection. Its operation is fixed by the installed nominal heating power and the fixed operation schedule throughout the year. Additional gaps are filled mostly by the heat pumps and by discharging the storage systems. The STC panels inject heat whenever they can, and surpluses are stored for later use. This is visible from the part below the zero power level, which indicates the charging behaviour of all tanks combined.

When the net power graph is studied in a shorter time range (see Figure 9), we can clearly see that power exceeding the neighbourhood heat demand is stored in the TES systems.

Around 40 GWh of TES is installed, and a clear seasonal charging pattern is obvious from the lower subplot in Figure 9. At the start of March, the storage tanks are completely empty, and during Spring and Summer they are gradually charged with energy until the maximum storage level is reached in November. Then, the storage is quickly depleted until the cycle repeats. A result of the genetic algorithms optimisation is that the full storage volume is used. Solutions where the charged energy profile does not use the full available range are dominated by more efficient designs.

An interesting result is that most storage tanks are used with a more or less similar charging pattern, witnessed by the evenly spread contributions of different systems. Only the smaller *ThorPark* TTES system is overshadowed by the other systems.

Figure 10 zooms in on a winter, spring and summer week to show a more detailed image of the heating power flows in the system. In the winter week, we see that the geothermal plant is working continuously, largely assisted by the heat pumps at their full power.

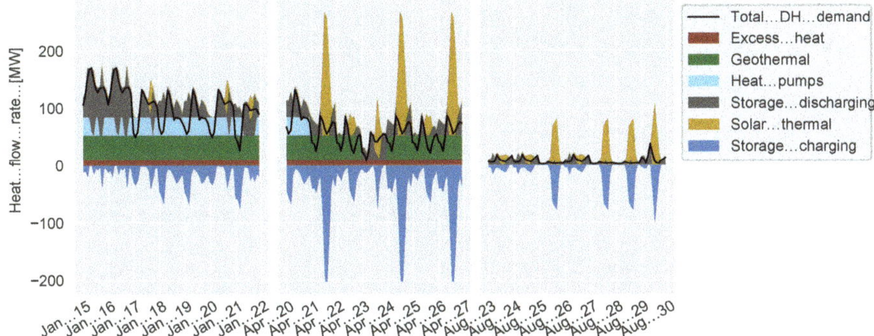

Figure 10. Detailed heating power plot of the minimum PEIS design for a winter, spring and summer week.

The remaining heat demand is met by discharging the TES systems. During the spring week, we see the intermittent behaviour of the geothermal plant (which is fixed by the off- and on-periods as determined by the user), but also the high power coming from the STC systems. Note that the TES tanks are mostly charging and only limited TES discharge is needed to fill some gaps in the heat demand. The heat pump is only on for higher demand levels. Note that the maximum amount of excess heat is being injected during both weeks (winter and spring). Finally, the summer week shows a very low heat demand, and while all "auxiliary" conversion systems are off, the demand is met by discharging the storage and power from the STC systems. Note how the surplus of solar heat is charged to the TES systems.

3.2.2. Intermediate LCOH and PEIS

The intermediate solution (see Figure 11) shows a rather different picture: whereas the previous solution was characterised by heat flow rate peaks over 200 MW, here the peak powers are more limited. The positive peaks do not exceed the maximum heat demand, which seems to suggest that the network has been designed to accommodate the maximum heat demand and not more. A slightly smaller geothermal system seems to be suggested by the power graph, but this is compensated by larger heat pumps. The lower positive peaks show that a smaller STC area has been installed.

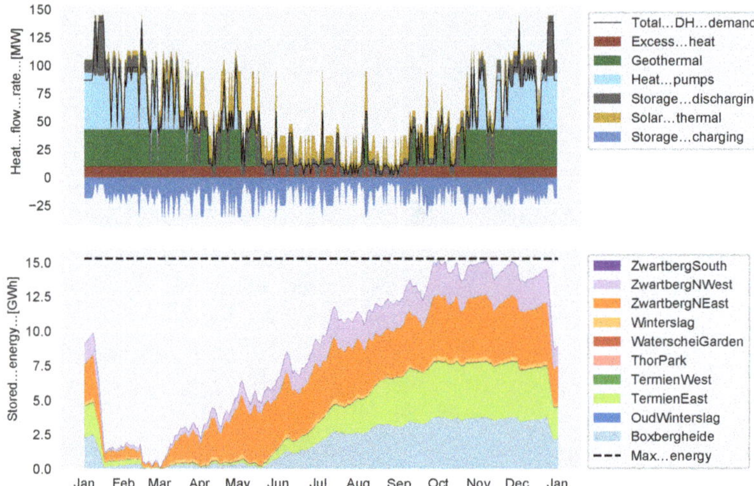

Figure 11. Full year time series for intermediate PEIS design.

The size of the TES systems is also considerably smaller, peaking at energy levels of 15 GWh. The seasonal behaviour is still apparent, but the charging behaviour is more gradual than that of the minimum PEIS solution. In addition, the faster charging and discharging cycles on top of the seasonal behaviour suggest that in this system, weekly and diurnal storage cycles have a more important role to play, contributing to the lower operational cost of this system. Finally, only four large storage systems are installed here, with the other TES systems playing only a marginal role. Further study will have to prove whether these TES systems will be completely removed if we leave the genetic algorithm to search even more generations with larger populations, or they are actually needed and worth investing in. Whereas the previous solution was characterised by similar charging patterns, the large storage tanks' storage behaviour is clearly shifted in time. To illustrate this, *Boxbergheide* and *TermienEast* only start charging in June, whereas *ZwartbergNEast* already starts filling up in March and remains at a more or less constant level from May until end of December.

Figure 12 again shows a more detailed picture of the heat flows. The behaviour looks largely the same as in Figure 10, the difference being in the scale of the graphs, which shows that the maximum power peaks are considerably lower than those in the minimum PEIS design. This is a result of the smaller STC systems installed here.

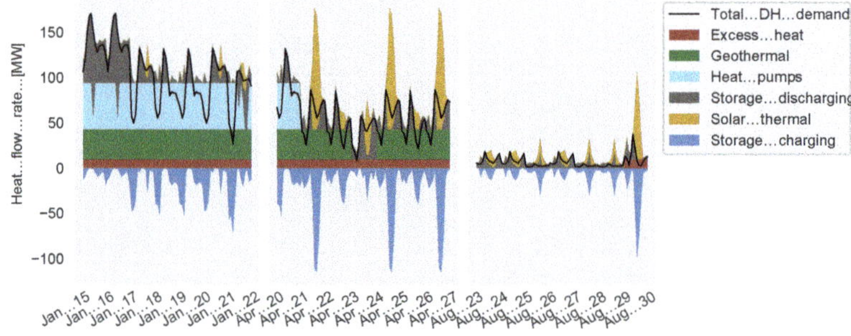

Figure 12. Detailed heating power plot of the intermediate PEIS design for a winter, spring and summer week.

3.2.3. Minimum LCOH, Maximum PEIS

The final highlighted design is that with a minimal LCOH (see Figure 13), but with the largest demand for primary energy from outside the network. Again, there are clear differences with regard to the previously discussed solutions: in this case, the excess heat injection is almost continuously at the maximum power level, whereas it was pushed out by solar power in the previous solutions. In this case, the installed STC area is very limited, which is witnessed by the absence of large positive power peaks.

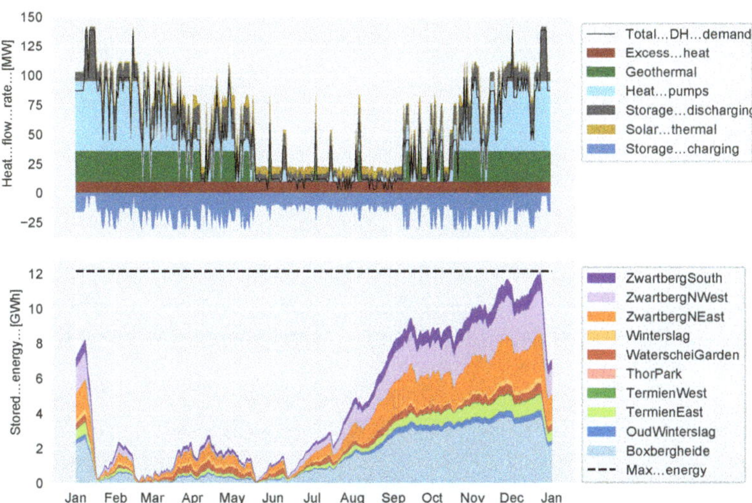

Figure 13. Full year time series for maximum PEIS design.

However, the TES volume is hardly smaller than that of the intermediate solution. It appears that a combination of the almost constant excess heat supply with the limited solar power injection makes it economically interesting to still have a decent amount of storage available in the network. In comparison to the intermediate solution, the average SoC is lower, which means smaller heat losses from the storage systems. Again, even though there is a clear seasonal charging pattern, there is an emphasis on shorter charging cycles to minimise operational costs.

Figure 14 shows the behaviour of the maximum PEIS system on a smaller time scale. Note the substantially lower solar thermal peaks compared to Figures 10 and 12. Another evident difference is the continuous injection of excess heat into the system, even during summer. Finally, we see that the heat pump is used more often during the spring week and even during the summer week.

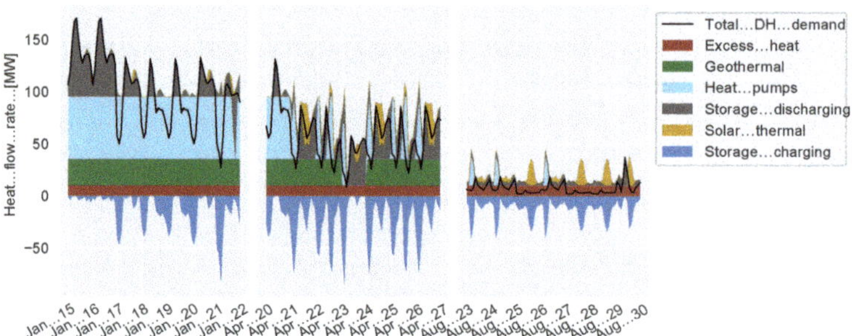

Figure 14. Detailed heating power plot of the maximum PEIS design for a winter, spring and summer week.

3.2.4. Comparison of Aggregated Results

To conclude this section on the three highlighted design solutions, Figure 15 summarises the heat delivery by the different heat sources in the network and compares these number to the total demand and storage losses. The difference between those two columns (sources vs. demand and losses) only differ by the amount of heat lost in the network. This figure makes clear that these network heat losses only make up a very limited fraction of the total heat balance. Although this result seems alarming at first, it can be explained by the fact that only the transmission network is modelled and the distribution network is omitted. Typically, these transmission pipes are much more efficient because of their wider diameter and the relatively low loss surface compared to the high mass flow rates that flow through these pipes. In addition, further analysis of the charging and discharging behaviour in these three highlighted solutions shows that an annual round-trip efficiency between 83% and 90% is achieved with the current storage model.

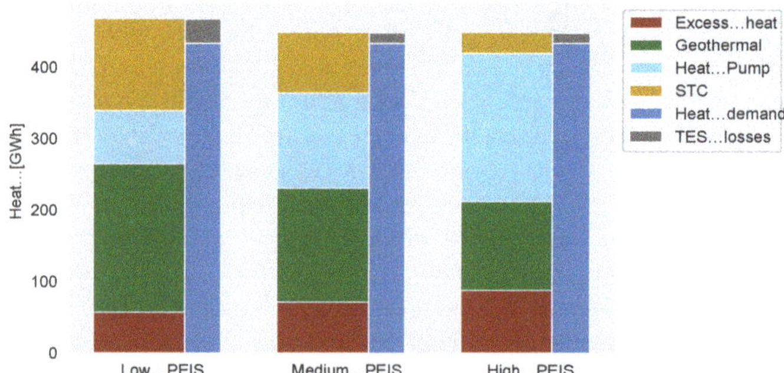

Figure 15. Summary of heat delivery for the three highlighted solutions.

3.3. *Influence of Network Temperatures*

This section compares the results of different temperature scenarios to the reference scenario with a 55/35 °C temperature regime. Figure 16 shows the differences between the resulting Pareto-optimal solutions with respect to the LCOH and PEIS objective functions.

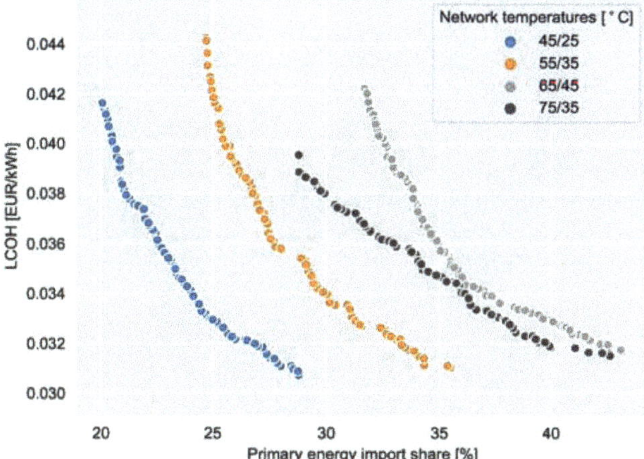

Figure 16. Comparison of the Pareto-optimal solutions for the different temperature scenarios. The excess heat cost is fixed at 15 EUR/MWh.

There is an obvious shift on the PEIS-axis, correlated with the average network temperature. This shift can most probably be mainly attributed to an increase in the COP (Coefficient of Performance) of the heat pumps and the geothermal plant with lower average network temperatures. The heat losses will also diminish with lower average temperatures, but given the limited influence of these losses, even more so with regard to the primary energy import into the network, it is expected that their role in the PEIS shift is of minor importance.

The 75/35 °C case is the only scenario with a 40 °C temperature difference, and this translates into a lower average LCOH. This is a result of the smaller pipes required for the same heating power transport compared to a system with a lower temperature difference, and by a more efficient utilisation of the same TES volume or STC area. The 20 °C ΔT scenarios are characterised by very similar LCOH ranges, but greatly differ in PEIS, where the reduced heat loss from the network and the increased COP of heat pump-based conversion systems are thought to be the main reason for these differences.

Considering the gas price in Belgium is currently between 0.036 and 0.039 EUR/kWh depending on the type of customer (Belgian Commission for Electricity and Gas Regulation CREG [35])—including the commodity price and the network cost, excluding taxes and levies, and disregarding the costs for a gas boiler or the loss of efficiency of a realistic heating system—we see that the current business as usual situation is already close to the LCOHs encountered in the future DES design. Compared to those numbers alone, already many of the proposed DES designs are actually cheaper than individual gas-based heating. At least, the system costs are clearly in the same order of magnitude. To make an honest comparison, the cost of the distribution networks (not considered in this study), but also the investment and maintenance of gas boilers would have to be included in the analysis. Hence, these results need to be interpreted carefully.

Figure 17 shows the Pareto fronts in terms of CO_2 intensity and LCOH for the different temperature scenarios. The carbon intensities for the network vary between 0.02 and 0.045 kg/kWh. For comparison, the specific CO_2 emissions for combustion of natural gas is 0.2 kg/kWh [36]. Even without accounting for the efficiency of a natural gas boiler, it can be seen that the CO_2 emissions of the studied designs are considerably lower. Clearly, Figure 17 is almost identical to Figure 16, except for the different x-axis. We refer to Appendix A.2.5 for a description of the time variation of the CO_2 intensity of the electricity consumption. Another important cost to include would be the current carbon tax. Moreover, this type of analysis could be used to determine which carbon tax levels would be needed to push the market towards more renewable technologies.

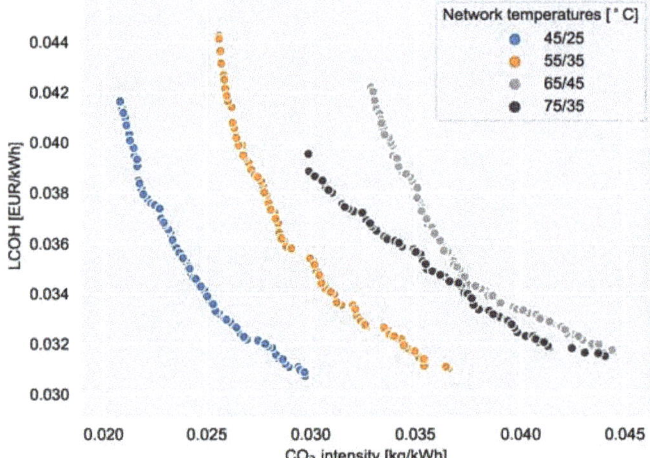

Figure 17. CO_2 intensity versus LCOH Pareto fronts for different temperature scenarios.

Figure 18 shows the relationships between the installed TES and STC sizes and their influence on LCOH and PEIS. In this figure, we see a more or less linear correlation for all temperature levels except the reference scenario. The 75/35 °C scenario has a slightly smaller TES volume distribution, but the spread on the STC area is more or less the same.

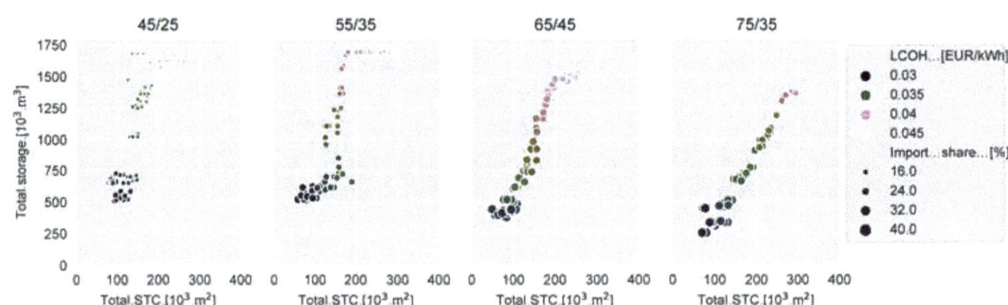

Figure 18. Correlation plot between total TES volume and STC area for GenkNet for different network temperatures. All plots have a fixed excess heat cost of 15 EUR/MWh.

3.4. *Influence of Cost of Excess Heat*

The differences in the Pareto fronts of scenarios with different excess heat cost—but with fixed network temperatures of 55/35 °C—are shown in Figure 19. The differences between the fronts are very subtle, and on the first sight they seem to be mostly coinciding. On closer inspection, the front with 20 EUR/MWh for excess heat is usually on top, indicating a slightly higher LCOH for the same PEIS, although in the higher LCOH range this is not always the case. In addition, the Pareto front of this highest excess heat cost stretches out further to the lower right than the other fronts. This means that lower cost solutions can be reached by importing more energy when the excess heat cost is higher, however the differences are very small.

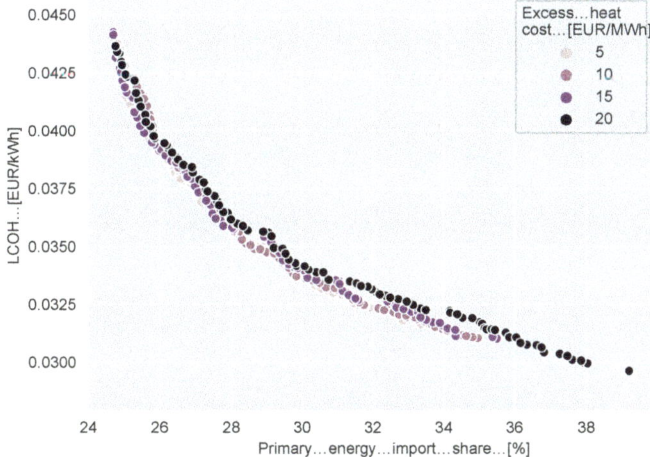

Figure 19. Comparison of the Pareto-optimal solutions for the different excess heat cost scenarios. The network temperatures are fixed at 55/35 °C.

These results suggest that the excess heat cost is of minor importance to the solution, at least for the cost range studied here. Indeed, the cost of excess heat is small compared to the total levelised costs encountered in the solution space. We expect to see larger differences when the excess heat cost approaches or exceeds the current system LCOHs. Moreover, the importance of the investment to connect the industrial excess heat sources is not to be forgotten. Because the availability of excess heat requires a district heating connection with a substantial investment cost, we expect that the point at which excess heat is no longer included in the optimal design will be already encountered at a lower excess heat price than the current system LCOH levels. Currently this tipping point at which excess heat is no longer chosen as a source of heat has not been encountered yet.

The main investment cost consideration influenced by the price of excess heat is the investment in the connection between the backbone and the node *GenkZuid* at which the excess heat is available. Therefore, we study the chosen diameter of this connection. Figure 20 shows the diameter for different energy ranges and excess heat costs, and it is clear that for the highest excess heat cost, the installed diameters are substantially smaller. Furthermore we see that the low energy import solutions always have a wider connection with the excess heat node than the ones with a higher energy range. Strangely, again we see an increasing trend in the diameters of the excess heat connection for the three lowest excess heat prices, which appears to be contradictory to the intuition that one would invest less to get access to a more expensive commodity, however also operational costs play.

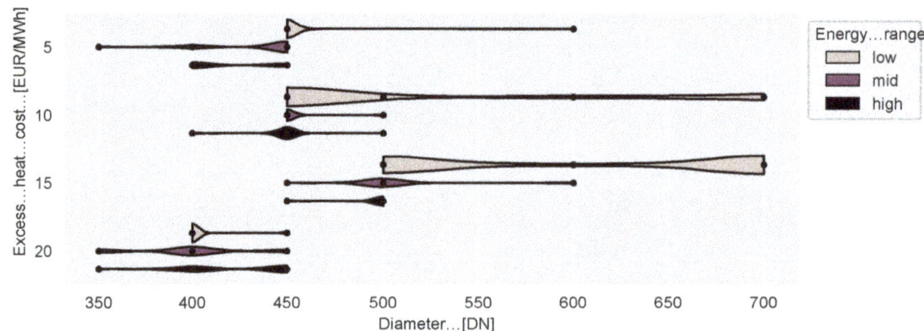

Figure 20. Influence of the excess heat cost on the diameter of the connection between the excess heat injection node and the rest of the network.

In summary, the cost of excess heat does not have a very clear influence on the design choices, except for the diameter of the excess heat node connection.

3.5. Calculation Times

For the seven calculated scenarios, the average calculation time for an entire genetic algorithm run (with 100 generations of 60 individuals) took 17.7 h. The algorithm was run on a Dell Precision 7920 Tower workstation with two Intel Xeon Silver 4116 processors (2.1 GHz–3.0 GHz Turbo, 12 cores) and 64 GB 2666 MHz DDR4 RAM. Hence, the average operational optimisation (a single evaluation) took 127.4 s, incorporating the fact that 12 processes were run simultaneously. Because the solver usually identifies an infeasible design much quicker, the average `modesto` calculation time for a feasible design is expected to be higher. However, a large spread on the calculation times was observed, where the problem was typically harder to solve as the system design got closer to its feasibility boundaries—that is, when the components were less oversized.

The fastest genetic algorithm run was performed in 12.0 h, the slowest in 24.9 h.

4. Discussion

The results have mostly shown that the genetic algorithm is able to select optimised designs using `modesto` as the evaluation core to calculate the objective function values and using representative days to reduce the calculation time. However, finding generalisable conclusions or rules of thumb based on these results is difficult. Even if we could find general rules based on the presented results, more data for different kinds of networks and additional scenarios would be needed in order to really generalise. Moreover, because of the heuristic nature of a genetic algorithm, we cannot prove the optimality of the results. Nonetheless, the convergence of the Pareto-optimal solutions is an indication that at least a local, and in the best case a global optimum is approached.

In addition, further study would be needed to increase the level of detail of the optimisation problem; for example, currently we don't account for the efficiency of substations that deliver the heat from the network to the end-users. Also the distribution networks in the separate neighbourhoods are not modelled at this moment; this means the investment is considerably underestimated, and also the heat demand might increase due to extra heat losses.

The omission of the building thermal flexibility from the model is expected to have a smaller influence, given the presence of seasonality in all storage SoC profiles, even for the solution with the highest share of primary energy import. Still, it would be interesting to see if the possibility of using the already available sources of energy flexibility (see Vandermeulen et al. [37,38] for a comprehensive discussion of this topic) in thermal networks would influence the design choices made.

One of the objectives of this thesis was to optimise the location of TES systems in a network. From the highlighted solutions and the distribution of the storage sizes, a possible conclusion would be that TES systems are preferably installed as close as possible to the heat sources in the network—that is, the heat pumps and geothermal plant. Of course, this result is biased since the STC systems also inject heat. But in general, the conclusion that heat should be stored as close as possible to where it will be generated is intuitive as it minimises transport losses.

The simultaneous optimisation of the network pipe diameters adds an interesting dimension to the results. Whereas smaller diameters are mentioned as one of the advantages of future, smart thermal networks (see Lund et al. [3]), the design results with maximal renewable and residual energy shares show that the pipe diameters actually become larger to enable large peak flow due to the redistribution of stored energy in the network and due to the high power injection by the solar thermal collectors.

A final remark must be made about the convergence of the genetic algorithm. The comparison of the solutions with different excess heat cost showed some counter-intuitive results, which could be explained by a possible incomplete convergence of the genetic algorithm. Upon closer investigation of the evolution of the hypervolume indicator (not included in the results of this paper), we see that the increase of the indicator flattens, indicating (near) convergence of the algorithm. Still, sometimes a sudden jump in the indicator—usually because of an accompanying jump in the Pareto front—is observed. Hence, it is possible that the most optimal Pareto front has not been reached yet. To be sure of the convergence, the genetic algorithm should be run with an even higher number of generations. On the other hand, the trade-off must be made between the optimality of the outcome, compared to the additional computation time needed to reach it. A similar trade-off is seen in the solution of MILPs, where an optimality gap is allowed to greatly reduce the solution time, at the expense of a slightly weaker certainty about the optimality of the solution.

5. Conclusions and Further Recommendations

This paper describes and illustrates an integrated design and control optimisation algorithm. For the evaluation of a single design, a Python toolbox named `modesto`, which implements a full year linear optimal control problem is used, in conjunction with an optimised representative days method to reduce the temporal complexity of the evaluations. By comparing designs using an optimal control problem, they can be compared objectively with regard to their maximum potential, and the comparison becomes independent of the efficiency of chosen control strategies. This optimal control problem is used as the evaluation core of a genetic algorithm, in which the design parameters are varied. As such, a two-layer design optimisation algorithm is constructed, where the lower layer optimises the operational aspect of the proposed networks, whereas the upper layer varies the design parameters in order to find the optimum with respect to a number of predefined objective functions.

This optimisation algorithm is applied to a fictive district heating network for the city of Genk in Belgium, and the influence of the network temperatures and the excess heat cost is investigated using a number of scenarios. It is shown that the design algorithm is able to efficiently find optimal solutions with respect to multiple simultaneous objectives, and that the proposed systems are competitive with individual natural gas-based heating systems in terms of levelised cost of heating, and even outperform this gas-based non-collective reference in terms of CO_2 emissions. However, we were unable to derive clear rules of thumb as to where the thermal energy storage must be installed. A clear result, though, is that seasonal energy storage will be crucial for future energy systems, as ample storage volumes are selected even for the cheapest solutions with the highest primary energy import into the network.

Suggestions for further research would be to add more modelling detail, for example by modelling the heat losses in the neighbourhood distribution networks explicitly (currently, only the transport pipes are modelled without considering the additional heat losses and investment in the neighbourhood distribution grids). Modelling the entire distribution network in detail however would probably result in very complex optimisation problems. Therefore, further study of how a homogeneous distribution network with distributed injection of renewable heat (STC panels on the rooftops of every building) could be aggregated, is required. Secondly, the `modesto` framework allows for easy addition of more energy conversion and thermal energy storage models. As such, the connection between different energy carriers (i.e., natural gas and electricity) could be strengthened with the addition of for example cobined heat and power (CHP) systems. This would also add a larger variation on the CO_2 objective, which is currently very strongly linked to the primary energy import into the network. As a last improvement to the optimisation toolbox, also the network temperatures could be varied, but in a deterministic way. The supply temperature could be varied as a function of the ambient temperature (heating curve), and as a pre-processing step this would maintain the linearity of the optimisation problem. However, this would require a modification of the energy storage models, which currently rely on fixed high and low temperature levels.

Author Contributions: Conceptualisation, B.v.d.H., A.V., R.S. and L.H.; methodology, B.v.d.H. and A.V.; software, B.v.d.H. and A.V.; verification, B.v.d.H.; writing—original draft preparation, B.v.d.H.; writing—review and editing, B.v.d.H., A.V., R.S. and L.H.; visualisation, B.v.d.H.; supervision, R.S. and L.H.

Funding: This research was funded by VITO grant numbers 1510475 and 1710760.

Acknowledgments: The authors want to acknowledge the work of Ina De Jaeger for the curation and analysis of the GIS database of Genk and for providing the building model parameters. We are grateful for the feedback on the results by Jan Diriken and Tijs Van Oevelen. Furthermore, we would like to thank Vincent Reinbold for the fruitful discussions that have contributed to this work. The work of Bram van der Heijde and Annelies Vandermeulen is funded through VITO PhD Scholarships.

Conflicts of Interest: The authors declare no conflict of interest.

Nomenclature

ΔT	Nominal network temperature difference [K]
$\dot{W}$	Mechanical power [W]
$d\dot{q}$	Heat loss rate per unit length [W/m]
$\dot{m}$	Mass flow rate [kg/s]
$\dot{Q}$	Heat flow rate [W]
τ	Economic lifetime
γ	Allowed variation on the nominal network temperature difference
A	Solar thermal collector area
$c_{maint,a}$	Annual maintenance cost
$c_{op,a}$	Annual operation cost
c_a	Total annualised cost
c_p	Specific heat of water [J/(kg K)]
i	Interest rate
I_{tot}	Total investment cost
I_a	Annualised investment cost
L	Pipe length [m]
R_s	Symmetrical thermal pipe resistance [Km/W]
T_a	Ambient temperature [K]
T_m	Average solar thermal collector panel temperature
T_r	Network return temperature [K]
T_v	Network supply temperature [K]
COP	Coefficient of Performance
DES	District Energy System
DHW	Domestic Hot Water
GHG	Greenhouse Gas Emissions
LCOH	Levelised Cost of Heating
MILP	Mixed Integer Linear Program
OCP	Optimal Control Problem
PEF	Primary Energy Factor
PEIS	Primary Energy Import Share
PSO	Particle Swarm Optimisation
PTES	Pit Thermal Energy Storage
R^2ES	Renewable and Residual Energy Sources
STC	Solar Thermal Collector
TES	Thermal Energy Storage
TTES	Tank Thermal Energy Storage

Appendix A. Operational Optimisation Models

This section provides a summary of model equations used. Because of the modular structure and the complex problems considered, giving a comprehensive overview of all optimisation variables and constraints is not possible.

Appendix A.1. Network Models

The heat losses from the district heating network are calculated (see van der Heijde et al. [39]) based on a nominal supply and return temperature. These nominal temperatures are also used by all other components in the DES, such as TES and conversion systems. To maintain the model's linearity on the one hand, and to avoid infinite temperature differences at low mass flow rates (in the case of nearly mass flow-independent heat losses) on the other, the heat losses from the pipes are made a linear function of the mass flow rate. The nominal heat loss level is reached around the mass flow rate that corresponds to a pressure gradient of 80–100 Pa/m.

Pumping losses in the network are modelled with a set of linear inequality constraints. These linear segments interpolate between equidistant points on the actual pumping power curve (third-degree function of mass flow rate). Because of the inequality constraints, only branched networks can be currently modelled, and the pumping energy must be represented in the operational objective function, such that high deviations from the inequality constraints are penalised. In this case, the (nonnegative) cost for the electricity to drive the pumps is a part of the operational cost.

As such, the following model equations can be written for a pipe model:

$$\dot{Q}_{in} = Ld\dot{q} + \dot{Q}_{out} \tag{A1}$$

$$\dot{m}c_p\Delta T \leq \dot{Q}_{in} \leq \dot{m}c_p(1+\gamma)\Delta T \tag{A2}$$

$$\dot{m}c_p\Delta T \leq \dot{Q}_{out} \leq \dot{m}c_p(1+\gamma)\Delta T, \tag{A3}$$

where γ denotes the allowed variation on the temperature difference in the network to allow heat losses without violating the energy balance. We limit γ to 5%, but the user can specify this parameter as needed. $\dot{Q}_{in}$ and $\dot{Q}_{out}$ are the respective heat flow rates at the inlet side and outlet side of the pipe system. ΔT is the design temperature difference between the supply and return side, neglecting any temperature differences between the in- and outlet side. L is the length of the modelled segment. Note that $\dot{Q}$ is defined positive when heat is transported from the inlet to the outlet of the pipe, and vice versa. Heat losses $d\dot{q}$ are positive when heat is lost from the pipe to the surrounding.

The linearised heat loss model is implemented as:

$$d\dot{q}(\dot{m}) = d\dot{q}_{nom}\frac{|\dot{m}|}{0.7233\dot{m}_{max}}, \tag{A4}$$

where

$$d\dot{q}_{nom} = \frac{T_v + T_r - 2T_a}{R_s}. \tag{A5}$$

The value of R_s is listed in Tables A2 and A3. 0.7233 is a factor that influences the slope of the heat loss curve, such that the nominal heat losses occur in the region around a pressure drop of 100 Pa/m.

Appendix A.2. Thermal Energy Conversion Components

The energy conversion units models are based on the Energy Hub concept (see Geidl et al. [40,41] or Evins et al. [42]). The conversion of one energy form to another is modelled using a predetermined efficiency factor. While the nominal network supply and return temperature to calculate transport heat losses are fixed, a variation on the temperature difference between supply and return γ is allowed to be able to account for heat losses in the network. This implies that also at the production side, the relation between the design temperature difference, mass flow and heat flow rates cannot be imposed strictly, to avoid non-linear equations.

Hence, the following inequality constraints on the mass flow rate and the temperature difference form a general representation of an energy conversion system in the thermal network:

$$\dot{m}c_p\Delta T \leq \dot{Q} \leq \dot{m}c_p(1+\gamma)\Delta T, \tag{A6}$$

where $\dot{m}$ is the mass flow rate injected into the district heating system's supply and extracted from the return side by the heat generating system and $\dot{Q}$ is the thermal power injected into the network by the conversion system. This mass flow rate is heated by at least ΔT, and at most $(1+\gamma)\cdot\Delta T$. These two inequalities make sure that the conversion unit is not able to inject heat into the network at zero mass flow rate, which would imply an infinite temperature difference. In addition, the maximum heat injection is limited to the nominal power of the conversion system, which is a design parameter:

$$\dot{Q} \leq \dot{Q}_{max} \tag{A7}$$

This means that even if a larger temperature difference than the nominal ΔT is used, the thermal output of the unit cannot be higher than its nominal power. There is no explicit bound on the mass flow rate, other than that following out of Equation (A6) and out of the mass flow rate limits of the pipes connected to the network node in which the conversion system is embedded. All conversion systems are assumed to be able to modulate their output 100%, unless stated differently.

Appendix A.2.1. Heat Pumps

The heat pumps considered in this study are large air source heat pumps. Its COP (Coefficient of Performance) is pre-calculated based on the variation of the ambient temperature, and based on the network temperatures:

$$\mathrm{COP} = \frac{\dot{Q}}{\dot{W}} = \eta_C \frac{T_v}{T_v - T_a}, \tag{A8}$$

assuming a non-ideal Carnot cycle with a relative efficiency $\eta_C = 0.6$ (compared to the ideal Carnot cycle). In this equation, $\dot{W}$ is the mechanical power supplied to the heat pump in the form of electricity, and T_v and T_a are the respective supply and ambient temperature. The Carnot efficiency, although seemingly optimistic, was chosen in line with data from the Danish Energy Agency [43] for large heat pumps.

Appendix A.2.2. Solar Thermal Collectors

The STC model assumes a steady-state heat delivery as a function of the mass flow rate and the solar irradiance only, and was derived from norm EN 12975-2 [44]:

$$\dot{Q}_{out}(t) = A \cdot \left(\eta_0 \dot{Q}_{sol}(t) - a_1\left(T_m - T_a(t)\right) - a_2\left(T_m - T_a(t)\right)^2\right) \tag{A9}$$

In this equation, $\dot{Q}_{out}$ and $\dot{Q}_{sol}$ represent the heat output and the solar irradiance on the unit surface respectively. The collector surface area is represented by A. η_0 is the base efficiency and a_1 and a_2 are temperature dependence parameters. Manufacturers of STC panels measure these values according to the norm mentioned before. This paper assumes Arcon Sunmark HT-SolarBoost 35/10 flat-plate collectors with an η_0 base efficiency of 0.839, an a_1 value of 2.46 W/(m^2 K) and an a_2 factor of 0.0197 W/(m^2 K^2) [45]. T_a is the ambient temperature, whereas T_m represents the mean panel temperature. We assume T_m to be the average of the supply and return temperature of the network, which would correspond with a linear temperature increase along the collector pipe. When the heat output would become negative according to Equation (A9), it is set to be exactly 0 W.

Solar panels are assumed to be oriented South at a 40° tilt angle. The solar panels studied in the design optimisation are either installed in an empty field in rows (no shading effects accounted for), or mounted on the south-oriented parts of the roof of the buildings in the selected neighbourhoods. The incident solar radiation on a tilted unit area was simulated in Dymola [46] using typical meteorological year data for Brussels. The effect of the incidence angle of solar radiation on the panel on the transmission and reflection of the solar irradiance is neglected for simplicity.

Appendix A.2.3. Geothermal Heating Plant

The Danish Energy Agency [43] considers a geothermal plant for district heating in combination with an electric heat pump that assists in extracting as much heat as possible from the ground. The water is pumped up from the ground extraction well and passes through a heat exchanger to preheat the water from the return side of the district heating system. The water in the geothermal loop is then cooled by the heat pump evaporator before it is pumped back into the injection well. The heat pump condenser in its turn injects heat into the district heating loop until to the desired supply temperature is reached.

A geothermal well is designed to be operated continuously and therefore it can hardly change the output power. We predetermine the period during which the geothermal plant is (in)active, usually in the summer months, in this case the plant is shut down from the end of May until the end of September (day 150 and 270 of the year).

The coefficient of performance of the system is determined based on the temperatures of the geothermal well doublet, combined with the design temperatures of the network and a pinch temperature difference of 5 K in the heat exchanger. This leads to a non-linear system of equations, which is solved as a pre-processing step and the resulting COP is used in `modesto`. We will not treat the non-linear system of equations in further detail here, as they can be derived from the system lay-out in Reference [43].

Appendix A.2.4. Industrial Excess Heat

Industrial excess heat is treated as a simple heat source with a fixed cost per unit of energy. It is assumed that the industry suppliers cover the entire investment and pumping costs and that these expenses are reflected in the excess heat price. Furthermore, we assume a constant availability of the nominal heating power, without the obligation to buy all of it. Down-periods for maintenance of the industrial processes are not accounted for.

Appendix A.2.5. Electricity Considerations

The electricity used by the heat pump components is assumed to be extracted from the electricity grid. We are using Belgian electricity grid data, namely the BELPEX day-ahead prices for the year 2014. The primary energy factor (PEF) and CO_2 intensity for the Belgian grid have been calculated by Vandermeulen and Vandeplas [47]. We have calculated the hourly average profile as a representation and assume the same profile is valid for every day of the year. The time series of these profiles (CO_2 intensity, PEF and electricity price) have been summarised in Figure A1. Note that the electricity price is incorporated in the representative days selection algorithm and hence the hourly average profile is not used in the optimisation framework. The reason for this choice is that the electricity cost has a large influence on the operational objective function of the optimisation, here the real yearly variation has been chosen instead of a daily aggregated version.

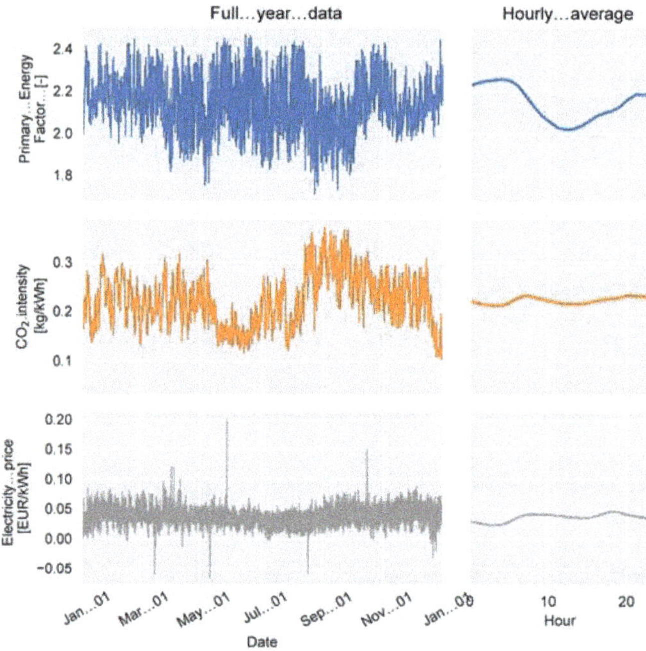

Figure A1. Time series and hourly average of PEF, CO_2 intensity and electricity cost.

The fact that these time series are not realistically correlated with the weather data is disregarded here.

Appendix A.3. Thermal Energy Storage

This paper considers two types of TES, namely tank and pit thermal energy storage (TTES and PTES). Both are modelled in line with the model used by Vandewalle and D'haeseleer [26], namely a perfectly stratified model (using the network supply and return temperature as the respective high and low TES temperatures), including state-dependent heat losses to the surroundings. The state of charge must be between 0% and 100%, and the (dis)charge heat flow is unconstrained.

Appendix A.3.1. Tank TES Systems

TTES systems have the advantage of better insulation over PTES systems, but due to their construction they are limited in size. In addition, the heat losses can be reduced even further by choosing an advantageous tank aspect ratio, minimising the tank surface with respect to the volume. We assume that the tank is constructed with a 0.5 m thick concrete shell, surrounded by 0.3 m of insulating material [48]. The concrete has a thermal conductivity of $1.63\,\mathrm{W/(m\,K)}$ and foam glass gravel is chosen as insulation, with a thermal conductivity of $0.095\,\mathrm{W/(m\,K)}$. Also, we assume that the wall and insulation thickness are the same for all of the tank walls. For TTES, we simply assume a cylindrical shape. Hence, the dependence of the heat losses is perfectly linear in the SoC. To minimise the surface area of the cylinder, we choose the aspect ratio $h/D = 1$.

Seasonal TTES systems are generally partly buried (or bermed) underground, which leads us to assume the average of the ambient and ground temperature as the boundary condition for the side walls of the tank. The bottom only sees the undisturbed ground temperature as a boundary condition, whereas the top of the tank is exposed to the ambient temperature.

Appendix A.3.2. Pit TES Systems

A comprehensive description about design of PTES systems can be found in Sørensen et al. [49]. Sorknæs [50] suggests a pit side and bottom wall conductivity of 0.5 W/(m·K) in case of sand, with an equivalent soil layer thickness of 2 m before the soil reaches the undisturbed ground temperature. The top insulation has a thickness of 0.24 m with a thermal conductivity of 0.104 W/(m·K). Because of the width variation at different depths, the relation between the SoC and the heat losses is no longer perfectly linear. However, it can be assumed to be linear without too large deviations. This is the result of the trapezoidal cross-section of the pit, where the base is narrower than the top edge. In this paper, a fixed shape of the storage pit is assumed, where the top width is 9 times the height, and the inclination of the sides is exactly 1 in 2 (or 26.57°), such that the bottom width becomes 5 times the height. Assuming a square top and bottom surface, the maximum considered pit volume of 200,000 m^3 has a height of approximately 16 m, whereas the top and bottom width are 144 m and 80 m respectively.

For the boundary temperature at the bottom of the pit, the undisturbed ground temperature T_g is used. The walls of the pit are assumed to be exposed to the average of the ground temperature and the outside air temperature, given the rather limited depth, including the part of the pit above the ground. The top cover is exposed to the ambient temperature only.

Appendix B. Techno-Economic Data

This section summarises data about costs, efficiency and lifetime of the used technologies.

Appendix B.1. Thermal Energy Conversion Systems

Table A1 summarises the used unit prices and investment analysis characteristics for the integrated optimal design and control algorithm. Note that cost figures per nominal power are always expressed considering the thermal power output.

Table A1. Characteristics of used conversion technologies.

Name	Investment	Fixed Maint. [% inv./y]	Lifetime [y]	Ref.
Heat pump	790 EUR/kW	0.60	20	[32,43]
Solar thermal collector	250 EUR/m^2	0.13	30	[32,51]
Geothermal heating	1600 EUR/kW	2.50	25	[52]

Appendix B.2. Thermal Energy Storage Systems

The unit cost of large storage systems varies with their size, hence the used cost data is represented graphically in Figure A2. The data points are derived from Schmidt and Miedaner [53]. The economic lifetime of TES systems is assumed to be 20 y, the fixed maintenance cost is 0.70 %/y with respect to the initial investment cost [32]. The economic lifetime and maintenance cost are the same for PTES and TTES.

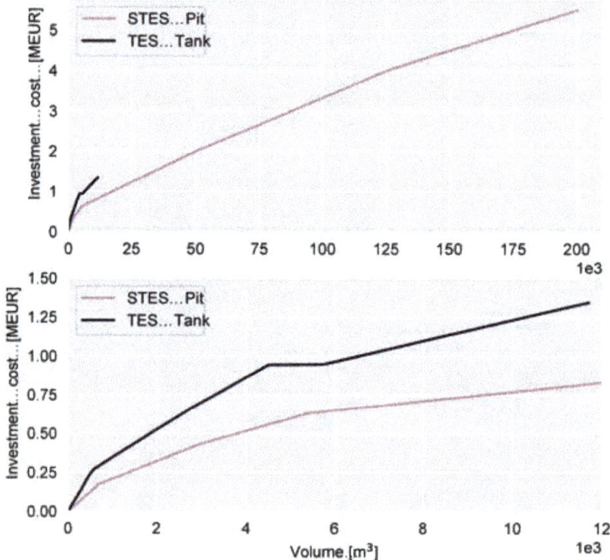

Figure A2. Cost evolution in function of volume for TES tanks and TES pits, using a linear interpolation between known investment costs for real systems. Data derived from Schmidt and Miedaner [53]. Note that the lower graph is representing the same data as the upper plot, but focussing on the lower volume range of the TES tank data.

Appendix B.3. Thermal Network Pipes

The available sizes of thermal network pipes and their dimensions are summarised in Tables A2 and A3. For the explanation of the different radii r and their respective accompanying diameters d, see Figure A3. The symmetric thermal resistance R_s is calculated using the derivation by Wallentén [54] and using the conclusions of van der Heijde et al. [39].

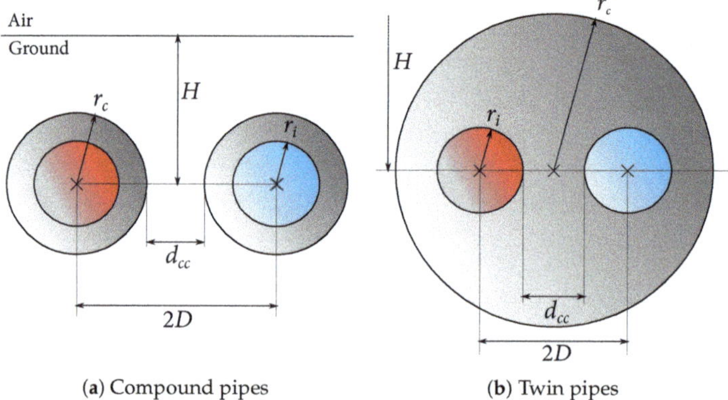

(**a**) Compound pipes (**b**) Twin pipes

Figure A3. Schematic representation of the types of double pipes considered in this study with clarification of the dimensions used to calculate the thermal resistance of compound and twin pipes. The grey shaded area indicates insulation material, the red and blue shaded areas represent the water in the supply and return pipes, respectively.

Table A2. Dimensions of twin pipes and resulting symmetrical thermal resistance. All distances are expressed in m, the thermal resistance per unit length R in K m/W. All data has been retrieved from Isoplus catalogues [33].

DN	d_c	d_i	d_o	d_{cc}	s	R_s
20	0.125	0.0217	0.0269	0.019	0.0026	12.259
25	0.140	0.0273	0.0337	0.019	0.0032	11.321
32	0.160	0.0360	0.0424	0.019	0.0032	10.208
40	0.160	0.0419	0.0483	0.019	0.0032	8.591
50	0.200	0.0539	0.0603	0.020	0.0032	8.623
65	0.225	0.0697	0.0761	0.020	0.0032	7.195
80	0.250	0.0825	0.0889	0.025	0.0032	6.277
100	0.315	0.1071	0.1143	0.025	0.0036	6.199
125	0.400	0.1325	0.1397	0.030	0.0036	6.629
150	0.450	0.1603	0.1683	0.040	0.0040	5.470
200	0.560	0.2101	0.2191	0.045	0.0045	4.878

d_o represents the outer diameter of the pipe wall, inside the insulation. d_{cc} is the distance between the walls of the two pipes, such that $D = \frac{d_{cc}+d_o}{2}$. s is the pipe wall thickness.

Table A3. Dimensions of compound single pipes and resulting symmetrical thermal resistance. All distances are expressed in m, the thermal resistance per unit length R in K m/W. All data has been retrieved from Isoplus catalogues [33].

DN	d_c	d_i	d_o	d_{cc}	s	R_s
250	0.40	0.2630	0.2730	0.4	0.0050	2.773
300	0.45	0.3127	0.3239	0.4	0.0056	2.437
350	0.50	0.3444	0.3556	0.4	0.0056	2.480
400	0.56	0.3938	0.4064	0.5	0.0063	2.340
450	0.63	0.4446	0.4572	0.5	0.0063	2.311
500	0.67	0.4954	0.5080	0.6	0.0063	2.024
600	0.80	0.5958	0.6100	0.7	0.0071	1.964
700	0.90	0.6950	0.7110	0.7	0.0080	1.744
800	1.00	0.7954	0.8130	0.8	0.0088	1.559
900	1.10	0.8940	0.9140	0.8	0.0100	1.426
1000	1.20	0.9940	1.0160	0.9	0.0110	1.306

For compound pipes, d_{cc} is the distance between the outer edge of the insulation jackets of the separate pipes, or $D = \frac{d_c+d_{cc}}{2}$.

In the study of Ahlgren et al. [55], a distinction is made between inner and outer city areas. Netterberg et al. [56] seem to have consulted a similar data source. They found a linear investment cost regression:

$$I_{inner} = 2.18 \cdot d_{DN} + 308.46 \ [\text{EUR/m}] \quad \text{(A10)}$$

$$I_{outer} = 1.8596 \cdot d_{DN} + 230.5 \ [\text{EUR/m}], \quad \text{(A11)}$$

where I_{inner} and I_{outer} denote the respective investment cost per unit length for inner and outer city areas and d_{DN} is the nominal diameter of the pipe in mm.

References

1. BPIE. *Europe's Buildings Under the Microscope*; Buildings Performance Institute Europe: Brussels, Belgium, 2011; ISBN 9789491143014.
2. European Commission. *Communication from the Commission to the European Parliament, the Council, the European Economic and Social Committee and the Committee of the Regions on an EU Strategy for Heating and Cooling*; Technical Report; European Commission: Brussels, Belgium, 2016.

3. Lund, H.; Werner, S.; Wiltshire, R.; Svendsen, S.; Eric, J.; Hvelplund, F.; Vad Matthiesen, B. 4th Generation District Heating (4GDH) Integrating smart thermal grids into future sustainable energy systems. *Energy* **2014**, *68*, 1–11, doi:10.1016/j.energy.2014.02.089. [CrossRef]
4. Dahash, A.; Ochs, F.; Janetti, M.B.; Streicher, W. Advances in seasonal thermal energy storage for solar district heating applications: A critical review on large-scale hot-water tank and pit thermal energy storage systems. *Appl. Energy* **2019**, *239*, 296–315, doi:10.1016/j.apenergy.2019.01.189. [CrossRef]
5. Paardekooper, S.A.; Lund, R.F.; Mathiesen, B.V.; Ojeda, M.A.C.; Petersen, U.R.; Grundahl, L.; David, A.; Dahlbæk, J.; Kapetanakis, I.A.; Lund, H.; et al. *Heat Roadmap Europe 4: Quantifying the Impact of Low-Carbon Heating and Cooling Roadmaps*; Technical Report; Aalborg Universitetsforlag: Copenhagen, Denmark, 2018.
6. Lund, H.; Østergaard, P.A.; Connolly, D.; Ridjan, I.; Mathiesen, B.V.; Hvelplund, F.; Thellufsen, J.Z.; Sorknæs, P. Energy storage and smart energy systems. *Int. J. Sustain. Energy Plan. Manag.* **2016**, *11*, 3–14, doi:10.5278/ijsepm.2016.11.2. [CrossRef]
7. Söderman, J.; Pettersson, F. Structural and operational optimisation of distributed energy systems. *Appl. Therm. Eng.* **2006**, *26*, 1400–1408, doi:10.1016/j.applthermaleng.2005.05.034. [CrossRef]
8. Söderman, J. Optimisation of structure and operation of district cooling networks in urban regions. *Appl. Therm. Eng.* **2007**, *27*, 2665–2676, doi:10.1016/j.applthermaleng.2007.05.004. [CrossRef]
9. Weber, C.I. Multi-Objective Design and Optimization of District Energy Systems Including Polygeneration Energy Conversion Technologies. Ph.D. Thesis, Ecole Polytechnique Fédérale de Lausanne (EPFL), Lausanne, Switzerland, 25 January, 2008; doi:10.5075/epfl-thesis-4018. [CrossRef]
10. Fazlollahi, S.; Becker, G.; Ashouri, A.; Maréchal, F. Multi-objective, multi-period optimization of district energy systems: IV—A case study. *Energy* **2015**, *84*, 365–381, doi:10.1016/j.energy.2015.03.003. [CrossRef]
11. Dorfner, J.; Hamacher, T. Large-scale district heating network optimization. *IEEE Trans. Smart Grid* **2014**, *5*, 1884–1891, doi:10.1109/TSG.2013.2295856. [CrossRef]
12. Morvaj, B.; Evins, R.; Carmeliet, J. Optimising urban energy systems: Simultaneous system sizing, operation and district heating network layout. *Energy* **2016**, *116*, 619–636, doi:10.1016/j.energy.2016.09.139. [CrossRef]
13. Falke, T.; Krengel, S.; Meinerzhagen, A.K.; Schnettler, A. Multi-objective optimization and simulation model for the design of distributed energy systems. *Appl. Energy* **2016**, *184*, 1508–1516, doi:10.1016/j.apenergy.2016.03.044. [CrossRef]
14. Patteeuw, D.; Helsen, L. Combined design and control optimization of residential heating systems in a smart-grid context. *Energy Build.* **2016**, *133*, 640–657, doi:10.1016/j.enbuild.2016.09.030. [CrossRef]
15. Lund, R.; Mohammadi, S. Choice of insulation standard for pipe networks in 4 th generation district heating systems. *Appl. Therm. Eng.* **2016**, *98*, 256–264, doi:10.1016/j.applthermaleng.2015.12.015. [CrossRef]
16. Prina, M.G.; Cozzini, M.; Garegnani, G.; Moser, D.; Oberegger, U.F.; Vaccaro, R.; Sparber, W. Smart energy systems applied at urban level: The case of the municipality of Bressanone-Brixen. *Int. J. Sustain. Energy Plan. Manag.* **2016**, *10*, 25–26, doi:10.5278/ijsepm.2016.10.3. [CrossRef]
17. Prina, M.G.; Lionetti, M.; Manzolini, G.; Sparber, W.; Moser, D. Transition pathways optimization methodology through EnergyPLAN software for long-term energy planning. *Appl. Energy* **2019**, *235*, 356–368, doi:10.1016/J.APENERGY.2018.10.099. [CrossRef]
18. Bornatico, R.; Pfeiffer, M.; Witzig, A.; Guzzella, L. Optimal sizing of a solar thermal building installation using particle swarm optimization. *Energy* **2012**, *41*, 31–37, doi:10.1016/j.energy.2011.05.026. [CrossRef]
19. Ghaem Sigarchian, S.; Orosz, M.S.; Hemond, H.F.; Malmquist, A. Optimum design of a hybrid PV–CSP–LPG microgrid with Particle Swarm Optimization technique. *Appl. Therm. Eng.* **2016**, *109*, 1031–1036, doi:10.1016/j.applthermaleng.2016.05.119. [CrossRef]
20. Vandermeulen, A.; van der Heijde, B.; Vanhoudt, D.; Salenbien, R.; Helsen, L.; Patteeuw, D.; Vanhoudt, D.; Salenbien, R.; Helsen, L. `modesto`—A multi-objective district energy systems toolbox for optimization. In Proceedings of the 5th International Solar District Heating Conference, Graz, Austria, 11–12 April 2018.
21. Remmen, P.; Lauster, M.; Mans, M.; Fuchs, M.; Osterhage, T.; Müller, D. TEASER: An open tool for urban energy modelling of building stocks. *J. Build. Perform. Simul.* **2018**, *11*, 84–98, doi:10.1080/19401493.2017.1283539. [CrossRef]
22. De Jaeger, I.; Reynders, G.; Saelens, D. Impact of spatial accuracy on district energy simulations. *Energy Proc.* **2017**, *132*, 561–566, doi:10.1016/j.egypro.2017.09.741. [CrossRef]

23. Baetens, R.; Saelens, D. Modelling uncertainty in district energy simulations by stochastic residential occupant behaviour. *J. Build. Perform. Simul.* **2016**, *9*, 431–447, doi:10.1080/19401493.2015.1070203. [CrossRef]
24. Hart, W.E.; Laird, C.D.; Watson, J.P.; Woodruff, D.L.; Hackebeil, G.A.; Nicholson, B.L.; Siirola, J.D. *Pyomo—Optimization Modeling in Python*, 2nd ed.; Springer Science & Business Media: Berlin, Germany, 2017; Volume 67, doi:10.1007/978-3-319-58821-6. [CrossRef]
25. Van der Heijde, B.; Vandermeulen, A.; Salenbien, R.; Helsen, L. Representative days selection for district energy system optimisation: A solar district heating system with seasonal storage. *Appl. Energy* **2019**, *248*, 79–94, doi:10.1016/j.apenergy.2019.04.030. [CrossRef]
26. Vandewalle, J.; D'haeseleer, W. The impact of small scale cogeneration on the gas demand at distribution level. *Energy Convers. Manag.* **2014**, *78*, 137–150, doi:10.1016/j.enconman.2013.10.005. [CrossRef]
27. Fortin, F.A.; De Rainville, F.M.; Gardner, M.A.; Parizeau, M.; Gagné, C. DEAP: Evolutionary algorithms made easy. *J. Mach. Learn. Res.* **2012**, *13*, 2171–2175, doi:10.1.1.413.6512. [CrossRef]
28. Deb, K.; Agrawal, S.; Pratap, A.; Meyarivan, T. A fast elitist non-dominated sorting genetic algorithm for multi-objective optimization: NSGA-II. In *International Conference on Parallel Problem Solving From Nature*; Springer: Berlin/Heidelberg, Germany, 2000; pp. 849–858.
29. Möller, B.; Werner, S. *Quantifying the Potential for District Heating and Cooling in EU Member States, the STRATEGO Project (Multi-Level Actions for Enhanced Heating)*; University of Flensburg: Flensburg, Germany, 2016.
30. Nussbaumer, T.; Thalmann, S. *Sensitivity of System Design on Heat Distribution Cost in District Heating*; Technical Report; International Energy Agency IEA Bioenergy Task, Swiss Federal Office of Energy & Verenum: Zürich, Switzerland, 2014.
31. Steinbach, J.; Staniaszek, D. *Discount Rates in Energy System Analysis*; Discussion Paper Commissioned by the Building Performance Institute Europe (BPIE); Technical Report May; Fraunhofer ISI and Buildings Performance Institute Europe (BPIE): Karlsruhe, Germany, 2015.
32. EnergyPLAN Modelling Team. *EnergyPLAN Cost Database*; Technical Report; Sustainable Energy Planning Research Group, Aalborg University: Aalborg, Denmark, 2018.
33. Isoplus. *Starre Verbundsysteme*; isoplus Fernwärmetechnik: Rosenheim, Germany, 2012. Available online: http://www.isoplus-pipes.com/download (accessed 20 October 2017).
34. Doračić, B.; Novosel, T.; Pukšec, T.; Duić, N. Evaluation of excess heat utilization in district heating systems by implementing levelized cost of excess heat. *Energies* **2018**, *11*, 575, doi:10.3390/en11030575. [CrossRef]
35. CREG. *A European Comparison of Electricity and Natural Gas Prices for Residential and Small Professional Consumers*; Technical Report; Commission for Electricity and Gas Regulation: Brussels, Belgium, June 2018.
36. IPCC. Emission Factor Database (EFDB). 2002. Available online: https://www.ipcc-nggip.iges.or.jp/EFDB/ (accessed on 7 April 2019).
37. Vandermeulen, A.; van der Heijde, B.; Helsen, L. Controlling district heating and cooling networks to unlock flexibility: A review. *Energy* **2018**, *151*, doi:10.1016/j.energy.2018.03.034. [CrossRef]
38. Vandermeulen, A.; Reynders, G.; van der Heijde, B.; Vanhoudt, D.; Salenbien, R.; Saelens, D.; Helsen, L. Sources of energy flexibility in district heating networks: Building thermal inertia versus thermal energy storage in the network pipes. In Proceedings of the Urban Energy Simulation Conference 2018, Glasgow, UK, 29–30 November 2018.
39. Van der Heijde, B.; Aertgeerts, A.; Helsen, L. Modelling steady-state thermal behaviour of double thermal network pipes. *Int. J. Therm. Sci.* **2017**, *117*, 316–327, doi:10.1016/j.ijthermalsci.2017.03.026. [CrossRef]
40. Geidl, M.; Koeppel, G.; Favre-Perrod, P.; Klockl, B.; Andersson, G.; Frohlich, K. Energy hubs for the future. *IEEE Power Energy Mag.* **2007**, *5*, 24–30, doi:10.1109/MPAE.2007.264850. [CrossRef]
41. Geidl, M.; Andersson, G. Optimal power flow of multiple energy carriers. *IEEE Trans. Power Syst.* **2007**, *22*, 145–155. [CrossRef]
42. Evins, R.; Orehounig, K.; Dorer, V.; Carmeliet, J. New formulations of the 'energy hub' model to address operational constraints. *Energy* **2014**, *73*, 387–398, doi:10.1016/j.energy.2014.06.029. [CrossRef]
43. Danish Energy Agency; ENERGINET. *Technology Data for Energy Plants Generation of Electricity and District Heating*; Technical Report; Danish Energy Agency: Copenhagen, Denmark, 2016.
44. European Committee for Standardization (CEN). *Thermal Solar Systems and Components—Solar Collectors—Part 2: Test Methods*; CEN: Brussels, Belgium, 2006.

45. Institut für Solartechnik. *SPF Online Kollektorkatalog*; Hochschule für Technik: Rapperswil, Switzerland, 2017. Available online: http://www.spf.ch/index.php?id=111 (accessed 30 January 2019).
46. Dynasim AB. *Dymola User's Manual*; Dynasim AB: Lund, Sweden, 2004.
47. Vandermeulen, A.; Vandeplas, L. Floor Heating in Residential Buildings: Optimisation Towards Different Objectives in a Smart Grid Context. Master's Thesis, KU Leuven, Leuven, Belgium, June 2016.
48. Guadalfajara, M.; Lozano, M.; Serra, L. Analysis of large thermal energy storage for solar district heating. In Proceedings of the Eurotherm Seminar 99-Advances in Thermal Energy Storage, Lleida, Spain, 28–30 May 2014; doi:10.13140/2.1.3857.6008. [CrossRef]
49. Sørensen, P.A.; Schmidt, T. Design and construction of large scale heat storages for district heating in Denmark. In Proceedings of the 14th International Conference on Energy Storage EnerSTOCK, Adana, Turkey, 25–28 April 2018; pp. 25–28.
50. Sorknæs, P. Simulation method for a pit seasonal thermal energy storage system with a heat pump in a district heating system. *Energy* **2018**, *152*, 533–538, doi:10.1016/j.energy.2018.03.152. [CrossRef]
51. Sørensen, P.A.; Nielsen, J.E.; Battisti, R.; Schmidt, T.; Trier, D. Solar District Heating Guidelines: Collection of Fact Sheets. Available online: https://www.solarthermalworld.org/sites/gstec/files/story/2015-04-03/sdh-wp3-d31-d32_august2012_0.pdf (accessed on 16 July 2019).
52. Danish Energy Agency. *Technology Data for Energy Storage*; Technical Report; Danish Energy Agency: Copenhagen, Denmark, 2018.
53. Schmidt, T.; Miedaner, O. *Solar District Heating Guidelines—Storage Components Fact Sheet*; Technical Report; Solites: Stuttgart, Germany, 2012.
54. Wallentén, P. Steady-State Heat Loss from Insulated Pipes. Ph.D. Thesis, Lund Institute of Technology, Lund, Sweden, 1991.
55. Ahlgren, E.O.; Simbolotti, G.; Tosato, G. *District Heating*; Technical Report; International Energy Agency (IEA): Paris, France, January 2013.
56. Netterberg, H.; Isaksson, I.; Werner, S. District Heating in Slough. Master's Thesis, Halmstad University, Halmstad, Sweden, 2009.

Article

An Approach to Study District Thermal Flexibility Using Generative Modeling from Existing Data

Camille Pajot [1,*,†], Nils Artiges [1,2,*,†], Benoit Delinchant [1,*], Simon Rouchier [2] and Frédéric Wurtz [1,*] and Yves Maréchal [1]

1 Univ. Grenoble Alpes, CNRS, Grenoble INP (Institute of Engineering Univ. Grenoble Alpes), G2Elab, F-38000 Grenoble, France; yves.marechal@g2elab.grenoble-inp.fr

2 University Savoie Mont-Blanc, LOCIE UMR CNRS 5271, Campus Scientifique SavoieTechnolac, F-73376 Le Bourget-du-Lac, France; simon.rouchier@univ-smb.fr

* Correspondence: camille.pajot1@g2elab.grenoble-inp.fr (C.P.); nils.artiges@g2elab.grenoble-inp.fr (N.A.); benoit.delinchant@g2elab.grenoble-inp.fr (B.D.); frederic.wurtz@g2elab.grenoble-inp.fr (F.W.)

† These authors contributed equally to this work.

Received: 30 July 2019; Accepted: 18 September 2019; Published: 24 September 2019

Abstract: Energy planning at the neighborhood level is a major development axis for the energy transition. This scale allows the pooling of production and storage equipment, as well as new possibilities for demand-side management such as flexibility. To manage this growing complexity, one needs two tools. The first concerns modeling, allowing exhaustive simulation analyses of buildings and their energy systems. The second concerns optimization, making it possible to decide on the sizing or control of energy systems. In this article, we analyze, in the case of an existing residential neighborhood, the ability to study by modeling and optimization tools two scenarios of energy flexibility of indoor heating. We propose in particular a method allowing to rely on a varied set of data available to build the various models necessary for optimization tools or dynamic simulation. A study was conducted to identify the neighborhood's flexibility potential in minimizing CO_2 emissions, through shared physical storage, or storage in the building envelope. The results of this optimization study were then compared to their application to the virtual neighborhood by simulation.

Keywords: district scale; demand-side management; flexibility; MILP; CO_2 emissions; heat pump; ETL; data management

1. Introduction

1.1. Energy Planning and Flexibility at the District Scale: Solutions and Issues

To fight climate change, many energy transition policies are emerging around the world [1]. With the ambition to achieve a successful transition from fossil fuels to low-carbon production, the share of renewable energies into the energy mix increases. It is well known that buildings represent more than a third of global energy consumption, 40% of CO_2 emissions, and much more in urban areas [2]. Besides, the integration of diverse renewable energy sources in cities is a major step to achieve sustainability objectives [3]. This diversity of solutions increases the complexity of urban planning, both for design and retrofit, when one has economical and energy efficiency in mind.

To cope with the energy landscape complexity, several works of research led to software developments towards energy planning. Theses energy planning tools target different time scales (time step and range) and space scales (local to global). Among them, we can cite for example:

- **MODEST:** The MODEST Energy System Optimisation Model aims to compute how energy demand should be covered at the lowest possible cost, using a model of energy networks suitable at regional and national scales [4]. MODEST uses linear programming (LP) to minimize capital and operation costs. The methodology uses a flexible time division to provide simulation results on both short and long time ranges.

- **OSeMOSYS:** Open Source Energy Modelling System is a generator of LP systems optimization models for long-term energy planning, from continent to village scale [5,6] with intra-annual resolution and 10–100-year time horizon. It relies on model blocks defining fuel inputs, regions, operative modes and usages, technologies, etc.
- **MESSAGE:** Model for Energy Supply Strategy Alternatives and their General Environmental Impact [7]. This LP model takes into account several energy generation technologies as well as carbon sequestration, with 5–10-year time step and up to 120 years of simulation range. It targets global and international scales.
- **TIMES:** The Integrated MARKAL-EFOM System (MARKet ALlocation-Energy Flow Optimization Model) is a LP/MIP (Mixed Integer Programming) model to evaluate several energy scenarios, combining a technical engineering approach and an economic approach, over medium- to long-term time horizons [8,9].
- **POLES:** Prospective Outlook on Long-term Energy Systems is a partial equilibrium energy and economic simulation model at the world scale [10,11]. It can model greenhouse gas emissions and final user demand as well as upstream production. It provides a yearly resolution and simulations up to 2050, with a Partial Equilibrium methodology.

All these models are great for testing and validating energy policies, energy landscape modifications at a wide scale, as well as studying medium- or long-term associated ecological and economic impacts. However, deep integration of intermittent renewable energies in the electrical network induces variability at the production side which could jeopardize the energy systems stability [12]. This phenomenon could be avoided by increasing the flexibility of consumption through demand-side management strategies, i.e., synchronizing the consumption with power production [13,14]. This area is more and more studied and especially applied to buildings whose consumption represents more than 55% of global electricity demand [2]. This raises a need for energy planning tools more suitable at a regional and medium scale (i.e., cities and districts). Many tools exist for this purpose. The reader can refer to the following reviews for an extensive overview: [15–17]. Among these, one can cite:

- **HOMER:** A commercial tool to help the design and the planning of micropower systems based on techno-economic analysis [18]. It provides simulation models with a minute resolution and several year time range.
- **REopt:** A commercial platform for energy planning with multiple technologies integration and techno-economic decision support [19].
- **Artelys Crystal Energy Planner:** A commercial software for the optimization and operational management of energy production assets in short- and medium-term [20].
- **Ehub Modeling Tool:** An open-source software package for preliminary design optimization of district energy systems based on Matlab [21].
- **DER-CAM:** A free decision support tool to help find optimal distributed energy resource investments [22]. Two main fields are investigated: buildings or multi-energy microgrids. It uses a MIP methodology, hourly and minute time step with up to 20-year time horizon.
- **Oemof-Solph:** A recent open-source modeling framework providing a toolbox to build energy systems models [23], with a MILP (Mixed-Integer Linear Programming) methodology and second to year time resolution.
- **Ficus:** An open-source software providing LP optimization models for capacity-expansion planning and unit commitment for local energy systems [24].

Generally, one can observe that such energy planning tools can be differentiated by the following criteria (see [25] for an extensive tools review):

- **Purpose:** The tool can be dedicated to the choice of investments, operation decisions or provide systems analysis.
- **Methodology:** All tools present simulation and/or optimization features. Many methodologies can be used. For optimization, LP (Linear Programming) is the most frequent methodology. MIP (Mixed Integer Programming) is also quite frequently used to handle discrete variables or Boolean.
- **Temporal resolution:** Each tool can feature a different time step range, from seconds to years.
- **Modeling temporal horizon:** As for temporal resolution, the maximum modeling temporal horizon is not the same for each tool and can go from a year to decades.
- **Geographical coverage:** According to the tool's objectives, geographical coverage ranges are different and can go from regional to global scales.

Besides, two types of strategies are mainly investigated for demand-side management: electrical appliances that can be shifted [26,27] and thermal loads that can be modulated (with or without storage system) [28]. The principle of heating load modulation without any storage system consists of using the internal mass of buildings as energy storage. Thus, a building can be over-heated when consumption is needed and under-heated when production is lower. To describe this behavior, Panão et al. introduced the concept of Building as Battery (BaB) and illustrated it on residential buildings with photovoltaic panels [29]. Although many studies only focus on the BaB, some others address the challenge of modulating the heating load with Thermal Energy Storage (TES) [30]. The evaluation of the flexibility is often realized thanks to simulation results [28,31].

Therefore, to tackle thermal flexibility at a district scale, the tool to use must be characterized by the following:

- Target system analysis for a good insight into technological choices and operation effects.
- Have MILP (Mixed-Integer Linear Programming) models and optimization strategy. Indeed, linearity is interesting for the scalability of optimization problems (and then convenient for city-scale studies). Furthermore, many systems present finite states (such as the storage system we use in this study), thus the optimizer must also support problems with integers.
- Provide dynamic thermal models of buildings and storage systems.
- Feature a time resolution compatible with building simulation models. Ten minutes is common in most building energy simulation software.
- Provide a regional geographical scale (district and city) with at least a decade of time horizon.
- Open sourcing can also be a relevant criterion since it fosters model and code sharing inside the community.

From our current knowledge, only Oemof-Solph and the open-source tool OMEGAlpes (Optimization ModEls Generation As Linear Programming for Energy Systems) we are developing in our team seem to comply with such requirements.

OMEGAlpes is dedicated to the generation of linear optimization problems for energy systems [32]. It allows quickly building Mixed-Integer Linear Programs (MILP) to design and manage multi-carrier energy systems. OMEGAlpes models are based on energy flows and energy units allowing to quickly study numerous cases by setting and gathering elementary models. Big optimization problems (hundreds of decision variables) can be quickly solved at the district scale due to linear models.

Oemof is more oriented towards interfaces between complementary tools and is currently less complete than OMEGAlpes on the model side, which led us to pursue our developments on OMEGAlpes for our studies on thermal flexibility in districts.

A final issue in the process of a flexibility study is the good choice of (building) models and their parameter values. This aspect is a key point in the field of Urban District Energy Modeling (UBEM) and is discussed below.

1.2. Objectives and Paper Structure

In this study, we aim to present a methodology to study the flexibility potential of a district that can be obtained by the heating modulation. In this study case, heating systems are decoupled from the domestic hot water because of different temperature levels. Thus, only heating load modulation is addressed. We show:

- How a MILP modeler such OMEGAlpes can be used to evaluate flexibility scenarios on a specific case.
- How one can use UBEM generation tools alongside existing data to produce the district MILP energy model.

The methodology is illustrated for a new residential district heated by a groundwater source. Located in Grenoble (France), the district is composed of 16 buildings outside-insulated with 11 floors on average. All buildings construction were initiated after 2010 and are designed according to the French energy policy for buildings (RT2012)—30% energy performance objective (30% more efficient than the RT2012, corresponding here to 50 kWh·m^{-2}·year^{-1} primary energy consumption). A simplified representation of the district is shown in Figure 1 and an overview in Figure 2. More details on geometric and physical parameters are given further in Section 2.3 Table 1.

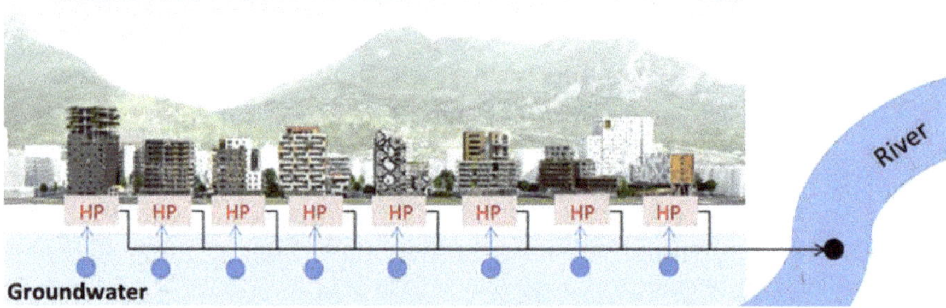

Figure 1. Illustration of the district thermal fed by a groundwater source.

Figure 2. Grenoble "Cambridge Eco-district" overview.

The goal of the study case is to quantify the reduction of CO_2 emissions that can be obtained through flexibility on the district heating load. First, we present how one can use existing data to produce a suitable district model in OMEGAlpes. Then, we describe models used in OMEGAlpes and how one can use this tool to study two thermal flexibility scenarios. The heating load modulation thanks to thermal energy storage is addressed and then compared to the Building as a Battery (BaB) concept. Finally, results obtained by optimization are confronted with simulation results obtained with simulation models.

2. Eco-District Modeling Based on Available Data

The first step in our methodology is to build a dynamic thermal model of our district suitable for MILP optimization. In this section, we explain why proper data management is important for district model generation and how a data management tool can be used alongside UBEM tools. Then, we describe the data used and the models generated for OMEGAlpes.

2.1. Data Variability and Heterogeneity

Contemporary cities, and particularly since the development of the concept of "smart cities", expose more and more data for various users and applications. The data exposition is fostered by various actors, with corresponding privacy levels. For example, one can access city-related data through the following sources:

- Open Data City Portal: Many cities expose today open data to their citizens through dedicated web platforms. For example, the city of Grenoble, France, delivers data on the portal [33]. The released data are highly dependent on the city's policy. One can find there infrastructure data, land registries, traffic and pollution metrics, monthly global electrical consumption, immigration rates, etc.
- Cartography data, from sources such as OpenStreetMap, satellite imagery, etc.
- Buildings databases, from general indicators [34] (EU buildings database: global aggregated indicators on buildings stocks, building construction dates, and consumption), to individual descriptions [35] (PSS—Archi: community-driven inventory of European buildings—addresses, GPS coordinates, height, usage, etc.), through statistical databases [36] (TABULA database, for statistical information about materials, usages, construction types, etc.).
- Energy certification/energy rating files: These files are dependent on a country's legislation, and generally produced before construction or during real estate transactions. For example, in France, the Thermal Regulation policy imposes the production of a "RSET" file for each new building containing structural, thermal and energy data used in a dedicated performance simulation software. They are most of the time produced by specialized engineering offices and not publicly available (one needs special inquiries to access them).
- BIM files (Building Information Models): These files are commonly created by engineering offices during building design. Similar to energy certification files, they are rarely freely available.
- On-site surveys: For specific projects, one can mandate surveys to recover buildings heights, number of floors, etc.
- Consumption data: Energy providers as customers have access to different aggregated consumption data according to standard privacy levels. Some data exchanges with energy providers are possible.

The main difficulties encountered by the engineer in obtaining city-scale data are as follows:

- Data accessibility and variability, due to the diversity of potential providers and inherent confidentiality policies.
- Data heterogeneity, due to the many various forms such data can hold.

Since data accessibility is more relevant to organizational problems and policies, the scientific challenge in exploiting these data is more related to their variability and heterogeneity. Indeed, four main axes of heterogeneity are observed: quantity, granularity, structure and semantics (see Figure 3). Therefore, to use these data for modeling purposes, one needs to apply appropriate tools and techniques to handle this heterogeneity.

The problem of data management is commonly encountered in the IT industry. In numerous application fields (online sales, social networks, advertising, etc.), developers have to handle various and sometimes unreliable data, encoded in different formats and databases. Generally, data are stored in different and specific databases (data warehouses) and not exploitable *as is*. One must then develop a middleware layer to extract the data, transform them and load them to client databases and interfaces to comply with clients' needs.

Our problem here is quite similar: we want to extract district-related data from various origins and encoded on various files, pre-treat these data and store them in models able to perform simulation and/or optimization tasks. Consequently, this approach of Extraction–Transformation–Loading (ETL) seems well adapted to the problem of district modeling from existing data. The reader can find general information about ETL techniques in [37].

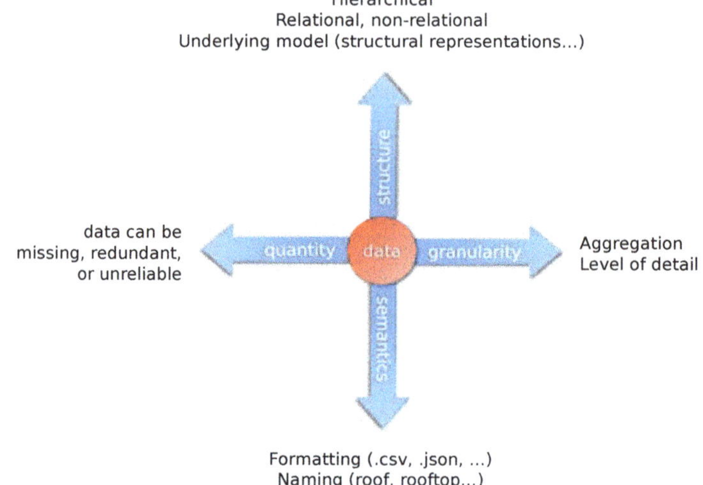

Figure 3. Data heterogeneity in districts.

2.2. District Buildings Modeling

Detailed dynamic thermal modeling of districts is more encountered in UBEM simulators tools such as CitySim [38], City Energy Analyst [39], TEASER [40] or CityBES [41], than in energy planning tools. The reader can refer to [42] for an extensive review. Considering the complexity to model a whole district, the amount of data potentially required and the nature of available data, some simplifying strategies are often used:

- Building thermal models simplification. Using low order RC models is a common approach.
- Definition of Archetypes/Prototypes models. Building types are categorized and a standard default model is defined for each category.
- Usage of BIM (Building Information Models) or dedicated city information models such as CityGML files.
- Individual parameters are often missing and then generated using statistical databases.

Among UBEM tools, TEASER is particularly adapted to the generation of low order models with few input data. This Python package developed at the University of Aachen can generate a simple "archetype" model of a building with a minimum of five parameters and can involve statistical databases for data enrichment. The generated models are Python objects translated in Modelica using IBPSA annex 60 or Aixlib libraries [40].

We build here "four walls elements (i.e., interior walls, exterior walls, floor plate, and roof) SingleFamilyDwelling" archetypes models using TEASER according to the IWU (Institut Wohnen und Umwelt—Institute for Housing and Environment) topology issued from the EPISCOPE project [36]. "SingleFamilyDwelling" corresponds to the archetype's data enrichment method. At the moment, TEASER mostly supports archetypes issued from studies of the German stock. This is not an issue here since we only test our methodology, but the implementation of French archetypes should be necessary for more accurate results. The resulting model is a RC reduced-order thermal model corresponding to the "AixLib.ThermalZones.ReducedOrder.RC.FourElements" component of the AixLib Modelica Library [43], as depicted in Figure 4. The corresponding Modelica simulation is further used as a reference for the virtual district.

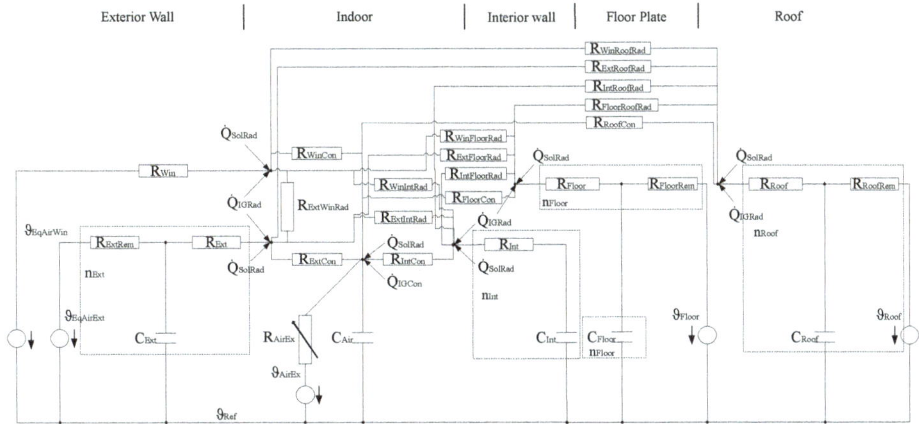

Figure 4. Four elements building model generated by TEASER [43].

A second model is required to perform optimizations. Thus, we developed a simpler linear building model in OMEGAlpes. Furthermore, too many model parameters can be counterproductive for an early-stage study. Therefore, we implemented a simplified RC-model of the Swiss SIA2044 norm in OMEGAlpes meeting our main needs: it can be easily built with few data, all the equations are linear (described Figure 5), and the Swiss building structure is very similar to the French one.

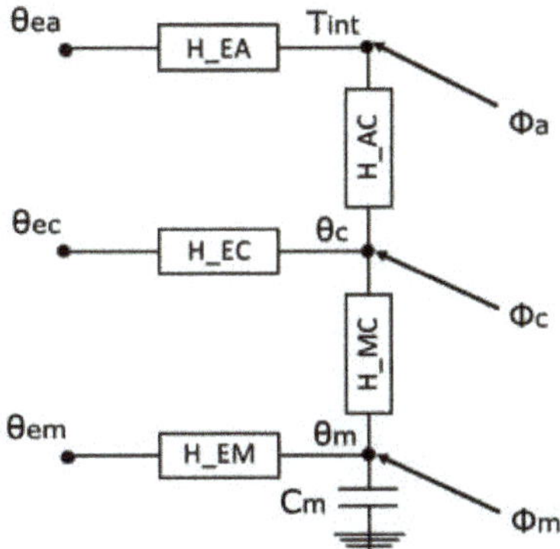

Figure 5. RC model according to SIA 2044.

This RC-model is composed of a thermal capacity C_m and five thermal resistances:

- H_EA: Heat transfer coefficient between the air node (a) and outdoor (e)
- H_EC: Heat transfer coefficient between the central node (c) and outdoor (e)
- H_EM: Heat transfer coefficient between the building mass node (m) and outdoor (e)
- H_AC: Heat transfer coefficient between the air node (a) and the central node (c)
- H_MC: Heat transfer coefficient between the building mass node (m) and the central node (c)

Parameters in TEASER models are translated to the OMEGAlpes model such that global thermal transfer coefficient (U) values and thermal capacitance are preserved.

2.3. Generation of the Building Stock Model from Existing Data

In our residential district study case, the following data are available:

- RSET files for eight buildings: These files stand for "Récapitulatif Standardisé d'Étude Thermique" (Standard Report of Thermal Study) and are mandatory in France for the construction of each new building since the application of the French thermal policy RT2012. Each of these files is an XML document containing relevant data such as U values, areas, structural information, HVAC devices description, and thermal performance coefficients.
- Grenoble city land registry: GeoJson file containing all building's footprints. This file is issued from the Grenoble Open data portal [33].
- A spreadsheet issued from engineering studies gathering general information on district buildings (addresses, dates of construction, heights, and number of floors).
- A meteorological file of one year of data.
- Various documentation issued from engineering offices involved in the district construction (electrical network map, heat pumps datasheets, etc.)

For OMEGAlpes building models, we need the following data:

- U values and areas of ground floors, roofs, walls and windows
- Absorbtivity and emissivity of walls, roofs, and windows
- Transmittance of windows
- Surfaces areas
- Net leased area
- Number of floors
- Building height
- Building type (small, medium or heavy)

All the data required for OMEGAlpes building models can be deduced from generated TEASER models. Then, the UBEM generation is processed according with the workflow summarized in Figure 6.

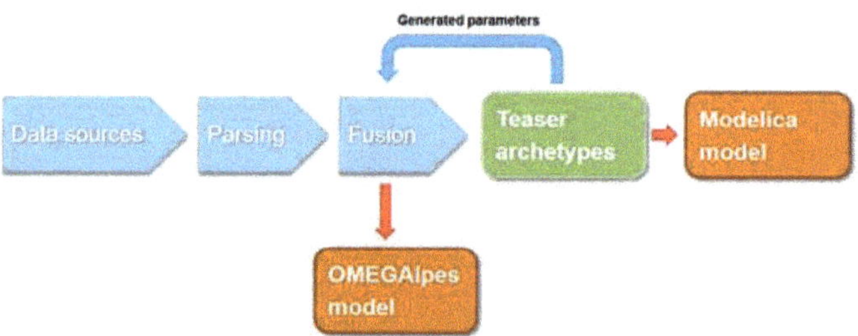

Figure 6. Workflow for district model generation.

In this workflow, one parses all the data files first. For buildings with RSET files, all the required parameters to build TEASER and OMEGAlpes models, except for emissivity, absorbtivity, and transmittance, are present. For other buildings, there are only enough data to generate TEASER archetypes. For some of them, the floor area is not directly available and one has to find corresponding polygons in the land registry file to estimate them. To complete missing data for OMEGAlpes models, one can extract parameters generated by TEASER in archetypes. For each parameter parsing, injection or extraction, one has to deal with different formulations and units. The generated dataset used to build all OMEGAlpes building models is summarized in Table 1.

Table 1. District main input parameters statistics for OMEGAlpes.

	Mean Value	Standard Deviation
Footprint/Roof area—m^2	322.6	119.7
Windows area—m^2	623	343.6
Ext. walls area—m^2	1875.9	993
U basement—$W{\cdot}m^{-2}{\cdot}K^{-1}$	0.76	*no variation*
U roof—$W{\cdot}m^{-2}{\cdot}K^{-1}$	0.17	0.025
U windows—$W{\cdot}m^{-2}{\cdot}K^{-1}$	1.61	0.31
U ext. walls—$W{\cdot}m^{-2}{\cdot}K^{-1}$	0.43	0.18
Walls emissivity	0.9	*no variation*
Roof emissivity	0.94	*no variation*
Windows emissivity	0.89	*no variation*
Walls absorbtivity	0.5	*no variation*
Roof absorbtivity	0.6	*no variation*
Windows transmissivity	1	*no variation*
Number of floors	10.56	3.01

To apply this workflow, we developed a specific Python package to ease file parsing, data manipulation (with an intermediate SQLite database) and district model generation, with modularity in mind (for further data integration). The architecture of this tool is summarized in Figure 7.

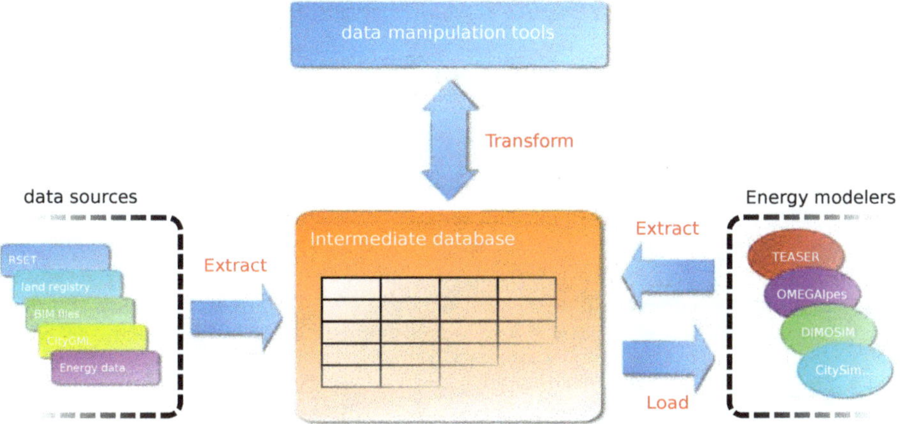

Figure 7. Python software architecture for data handling and model generation.

Such an approach is close to the ETL methodology. As ETL processes are well suitable for UML modeling [44], the choice of an Object-Oriented Programming language such as Python is appropriate. Besides, the support of Python by the scientific community eases the development of the "Transform" part.

3. Modeling of the Optimal Planning Problem

3.1. Optimal Planning of the District Heating Systems with OMEGAlpes

As already mentioned, two flexibility approaches to manage the district heating systems are addressed through the study case:

- The first one consists in designing an energy system composed of a heat pump and thermal energy storage to minimize the CO_2 emissions of a fixed district heating load.

- The second study case also aims to minimize the CO_2 emissions of the buildings' heating load, but thanks to flexibility through building envelopes. In this case, specific building models dedicated to the optimization should be used to estimate how the load can be modulated.

In both studies, we aim to estimate the possibility to decrease the CO_2 emissions by designing and operating the system. The studies were conducted during two weeks in January, which usually represent the coldest period and are critical for the power system. Thus, the design of the system can be significant for the entire heating period. It is important to notice here that our goal is not to predict energy needs and CO_2 emissions for an entire year, but to be closer to the operation. Therefore, focusing on two weeks allows us to anticipate the possibility to pursue our work with a model predictive control approach thereafter.

To define the OMEGAlpes optimization models, a graphical formalism was defined to represent the energy units and power flows (see Figure 8).

Heat	Electricity
Production	Production
Consumption	Consumption
Storage	Storage

	Heat	Electricity
Power flow	→	→
Energy node	○	○

Figure 8. OMEGAlpes formalism for energy system modeling.

Let us introduce Study Cases 1 and 2; the results are detailed in Section 4.

3.2. Study Case 1: Flexibility through Thermal Energy Storage (TES)

The first study case deals with energy flexibility provided by a Thermal Energy Storage (TES) to minimize the CO_2 emissions of the district heating load. The energy system studied is composed of the district heated by geothermal groundwater through a heat pump and thermal energy storage to provide demand-side management. The goal of this study case is to design the whole supplying system (heat pump and storage). To do so, we used three OMEGAlpes units to model the energy system: the district heating load, the heat pump, and the thermal energy storage, as shown in Figure 9.

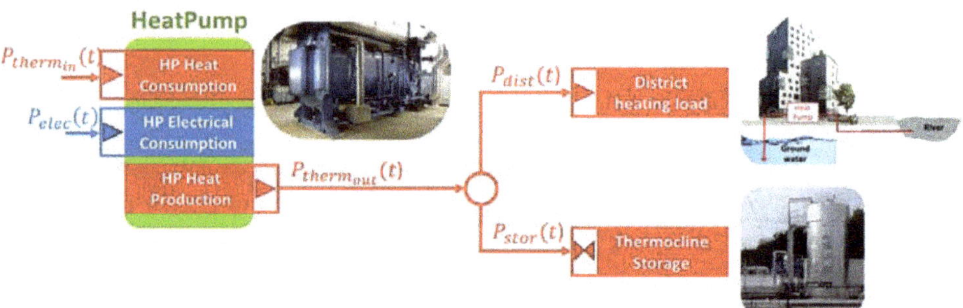

Figure 9. Modeling of the energy system of the first study case (heat pump, district heating load and thermal energy storage) according to OMEGAlpes formalism.

3.2.1. Estimation of the District Heating Load

In this first study case, only the flexibility provided by the storage energy system is addressed so that the district heating load cannot be modulated and is thus an input of our optimization problem. To estimate the thermal needs of the district, we relied on results from a first optimization obtained with OMEGAlpes which can be considered as a temperature regulation simulation. All buildings were modeled as described in the previous section and set with standard occupancy schedules obtained by TEASER and a temperature set-point of 20 °C. The objective of the optimization is to minimize the sum of the over-heating and the result (see Figure 10) is taken as the dynamic thermal consumption of the district P$_{dist}$(t). In this figure, we can notice that, during the days, the district heating load is very low. This could be explained by the high insulation of the buildings which require low consumption and thus can benefit from occupancy and solar gains to cover their needs.

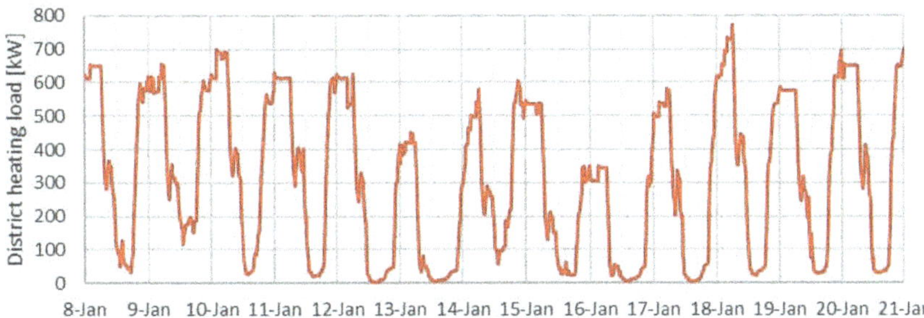

Figure 10. District heating load during a two-week period in January obtained by optimization.

3.2.2. Modeling of Heat Pump

Composed of new residential buildings, the district can be heated by low temperatures around 35 °C. With a groundwater temperature around 15 °C, this specificity allows the heat pump to reach a high Coefficient Of Performance (COP) of 5. Moreover, the temperature of the groundwater is assumed to be invariant so that we can consider the COP to be a constant equal to 5. Therefore, the heat pump is modeled by the relations between the thermal power provided by the groundwater $P_{therm_{in}}(t)$, the electrical power consumed by the heat pump $P_{elec}(t)$, the thermal power delivered $P_{therm_{out}}(t)$ and the COP, as described in Equation (1).

$$\begin{cases} P_{therm_{in}}(t) + P_{elec}(t) = P_{therm_{out}}(t) \\ P_{therm_{out}}(t) = COP * P_{elec}(t) \qquad \text{Where: } COP = 5 \end{cases} \tag{1}$$

In this study case, a trade-off was chosen between different levels of accuracy of the whole energy system modeling according to the uncertainties relating to the occupants' behaviors. Indeed, as we aim to estimate orders of magnitude of the CO_2 emissions reduction obtained by heating flexibility, the modeling of the heating systems is very simplified. For further studies, a deeper level of modeling could be needed to provide a more accurate estimation.

3.2.3. Modeling of Thermal Energy Storage (TES)

Multiple types of thermal energy storage systems are used in the literature to smooth building thermal needs. However, the most widespread technology used remains water tanks for their simplicity and low costs.

The power stored to the TES $P_{stor}(t)$ is defined as the difference between the charging power $P_c(t)$ and the discharging power $P_d(t)$ as described by Equation (2).

$$\begin{cases} P_{stor}(t) = P_c(t) - P_d(t) \\ P_c(t) \geq 0 \\ P_d(t) \geq 0 \end{cases} \tag{2}$$

Moreover, the relation between the energy contained in the water tank $e(t)$ and the charging and discharging powers is defined by Equation (3). The storage capacity C_{stor} is defined as the maximal value of $e(t)$.

$$\begin{cases} e(t+dt) = e(t) * (1-\alpha_{sd}) + (P_c * \eta_c - \frac{P_d}{\eta_d}) * dt \\ e(t_0) = e(t_f) \\ e(t) \leq C_{stor} \end{cases} \quad (3)$$

where

- α_{sd} is the coefficient of self-discharge of the storage system (depending on the storage design). Here, the coefficient is a percentage per time step (dt = 10 min).
- η_c/η_d is the charging/discharging efficiency (standard value of 95% corresponding to actual TES).
- t_0/t_f is the starting/ending time step of the period.
- dt is the time step (10 min).

In this study, a stratified storage system is considered called thermocline storage whose management is more complex than traditional storage (more details can be found in [45]). Indeed, in our case, we assumed that the storage has to be fully charged at least once per five days to optimally operate. The first step to model this constraint is to define a variable to indicate if the storage is fully charged. To do so, a binary variable was introduced: *is_soc_max(t)* which equals 0 when the state of charge is lower than 100% and 1 when the storage is fully charged. The definition of this indicator was realized thanks to Equation (4), where C_{stor} is the storage capacity, $e_{stor}(t)$ is the energy contained in the storage at the time t and ϵ is taken equal to 10^{-3}.

$$\begin{cases} C_{stor} * (1 + is\_soc\_max(t) - \epsilon) \geq e_{stor}(t) \\ C_{stor} * is\_soc\_max(t) \leq e_{stor}(t) \end{cases} \quad (4)$$

Then, our constraint can be expressed thanks to a sliding window including five days. Let t_{cycl} be the time step corresponding to the end of the first five-day period; the constraint of at least one full charge during five days is defined by Equation (5).

$$\forall t \in [t_{cycl}; t_f], \sum_{k=t-t_{cycl}}^{t} is\_soc\_max(k) \leq 1 \quad (5)$$

3.2.4. Modeling of CO_2 Emissions of the District Heating Load

In this study case, the CO_2 emissions of the district heating load (Em_{CO_2}) come from the electrical consumption of the heat pump. Fed by the French power system, the heat pump emissions vary dynamically according to the French grid CO_2 emissions rate ($em_{CO_2,rate}(t)$, see Figure 11).

Thus, the CO_2 emissions of the district heating load can be calculated by Equation (6), so that changing the heat pump operation could lead to CO_2 reduction, which we tried to achieve thanks to thermal energy storage in this study case.

$$Em_{CO_2} = \sum_{t=t_0}^{t=t_f} P_{elec}(t) * em_{CO_2,rate}(t) \quad (6)$$

Figure 11. CO_2 emissions rate of the French power system during a two-week period in January 2018.

3.2.5. Energy System Design Parameters

As explained above, the objective is to minimize the CO_2 emissions of the district heating load. To do so, we considered a groundwater source heat pump coupled with the thermal energy storage that we aim to design. Three parameters are optimized:

- The storage capacity (C_{stor}): Increasing the storage capacity allows more energy to be stored and thus the possibility to provide the thermal needs with the TES during high-CO_2 periods. However, big storage capacities induce higher costs and volume. In this study, we considered TES with capacity from 100 kWh to 48 MWh.
- The storage insulation, defined by the self-discharge coefficient (α_{sd}): An important factor in the storage design is the possibility to shift the energy in the medium term (several hours to days). This essentially depends on the self-discharge coefficient. If it is too high, too many losses will appear and it would be less efficient to shift the energy in the medium-term. In this study, we compared the influence of three values of α_{sd}: 0.125%, 0.25% and 0.5%, each ten minutes.
- The maximal electrical power consumed by the heat pump (P_{elec}^{max}): Increasing the power that can be consumed by the heat pump leads to higher thermal power delivered at a low-CO_2 period. Nevertheless, it induces high consumption peaks that are usually harmful to the power grid. In this study, we went from no over-sizing of the heat pump (300 kW) to 2500 kW.

3.3. Study Case 2: Flexibility through Heating Loads Modulation (BaB)

In this second study case, thermal flexibility is provided by the building envelopes. Each building is modeled individually so that the district load can be deduced by the addition of each building heating load. Thus, the thermal load of the district can be directly modulated without any external thermal energy storage (see Figure 12).

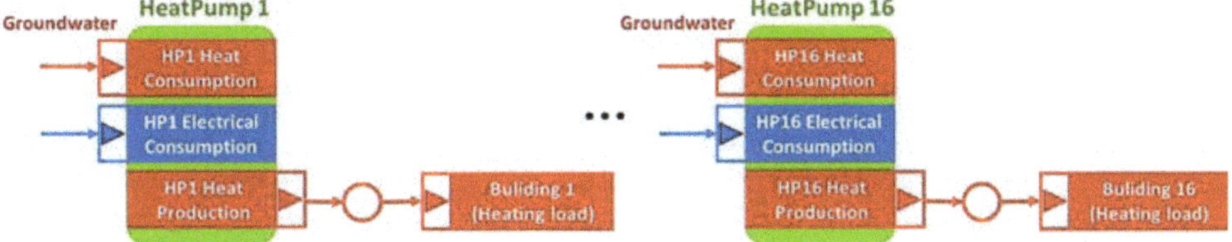

Figure 12. Modeling of the energy system of the second study case (16 buildings and 16 heat pumps) according to OMEGAlpes formalism.

3.3.1. Estimation of the District Heating Load

To guarantee the occupants' thermal comfort, the operative temperature is constrained to be higher or equal to 20 °C. Thus, the building can be over-heated by moments to store heat into the buildings while keeping thermal comfort.

The heating load can be calculated for each time step according to the thermal RC model available in OMEGAlpes and presented in the previous section. Besides constraining the operative temperature, the boundaries conditions are the same as before. These internal gains (from occupancy and weather) are applied to the nodes a, c and m (see Figure 5). More details about the model can be found in [46].

3.3.2. Modeling of Heat Pumps and CO_2 Emissions of the District Heating Load

The configuration of the district is slightly changed since each building is fed by its heat pump. Each heat pump is designed with an over-sizing (around +66%) according to the reference heating need of being able to use flexibility. The total maximal electrical consumption allowed to feed all the heat pumps was set to 500 kW.

The modeling of the CO_2 emissions is similar to the previous study so that each heat pump emits according to its electrical consumption. However, the objective to minimize the CO_2 emissions is global.

In this study case, the energy system is designed before running the optimization. Therefore, the minimization of the CO_2 emissions is based on finding an optimal operation of all the heat pumps of the districts.

4. Results

This section is divided into two main subsections:

- Optimization: Presentation of the optimization results for the two study cases aiming to reduce the CO_2 emissions of the district heating load. Here, reduced building models are used to predict heating thermal needs.
- Simulation: A reference scenario is compared to the simulation results obtained by setting the temperature profile according to optimization results with flexibility.

4.1. *Optimization*

The study cases presented in this paper are realized for a time step of 10 min for two weeks in January. For the first one, each optimization problem generated is composed of 38k variables (28k continuous and 10k binaries) for 61k constraints. The resolution was launched on an Intel bicore i5 2.4 GHz CPU with the Gurobi solver so that the optimization problem was solved within less than 10 s on average for 192 optimizations. The corresponding results are detailed Section 4.1.2.

The second study case consists of a single resolution since only one configuration is studied. The associated optimization problem is composed of 1211k variables (1100k continuous and 111k binaries) for 1263k constraints. The resolution was launched on the same Intel bicore i5 2.4 GHz CPU with the Gurobi solver and the optimization problem was solved within 23 min. The dynamic results are detailed Section 4.1.3.

4.1.1. Flexibility Potential

To evaluate the gains obtained by optimization, the first step is to estimate the maximal reduction in CO_2 emissions that can be achieved. A simple way to evaluate this maximum is to allow shifting each 10-min power slot of the load curve to minimize the CO_2 while consuming the same energy during the two weeks. In this case, the CO_2 emissions can be reduced to a maximum of 22% keeping the current heat pump maximal power consumption (300 kW). Of course, in this case, the building comfort (internal temperature) is not guaranteed. This potential of 22% CO_2 savings is used to compare the next results.

The dynamic result of this naive estimation is shown Figure 13. We can notice that some CO_2 emissions reduction can be obtained by anticipating or removing the consumption for a few hours (short-term flexibility), but the longer-term variation of the CO_2 levels lead to a need for longer-term flexibility.

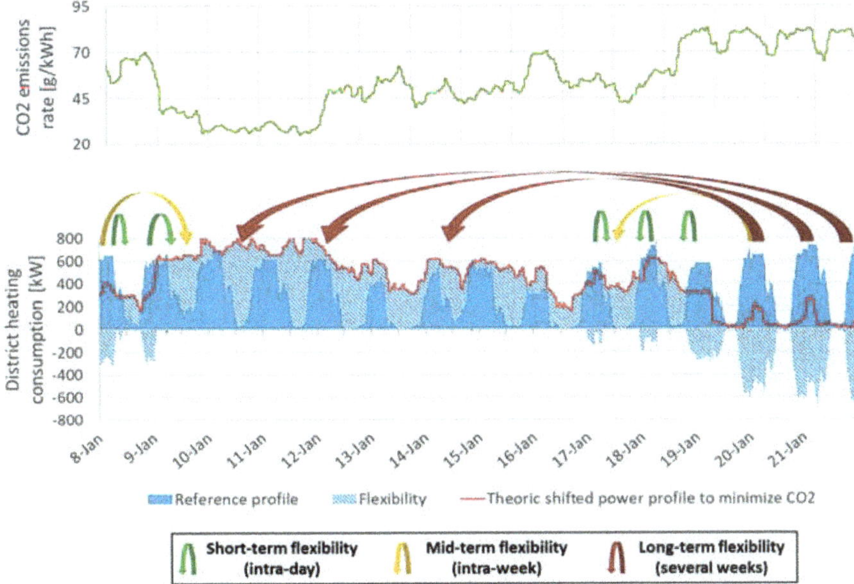

Figure 13. Theoretical maximal flexibility to minimize CO_2 emissions by shifting 10-min values from the reference profile.

4.1.2. Study Case 1: Flexibility through Thermal Energy Storage (TES)

In this study case, three elements were designed: the storage capacity (C_{stor}), the storage self-discharge coefficient (α_{sd}) and the maximal electrical power consumed by the heat pump (P_{elec}^{max}). Results are shown in Figure 14 for the three self-discharge coefficients studied (0.125%, 0.25% and 0.5%). The CO_2 emissions reduction obtained in each configuration is drawn according to the storage capacity and the maximal electrical power consumed by the heat pump.

Regarding the storage capacity, we can notice that 100 kWh is too small to reduce the CO_2 emissions regardless of the two other parameters. For larger capacities ($\geq$1 MWh), the impact on the CO_2 emissions reduction begins to be noticeable and is strongly correlated with the self-discharge coefficient and the maximal power consumed by the heat pump.

For a storage capacity under 2 MWh, we can notice that the reduction in CO_2 emissions are lower than 3.5% for all designs. However, in the case of a TES of 48 MWh with 0.125% of self-discharge and a maximal electrical power of the heat pump of 2500 MW, we manage to reach 20% reduction in CO_2 emissions of the district heating load. Knowing that the average daily heating consumption of the district during the period is 8 MWh, a 4 MWh storage corresponds to 12 h of consumption while a capacity of 48 MWh corresponds to six days.

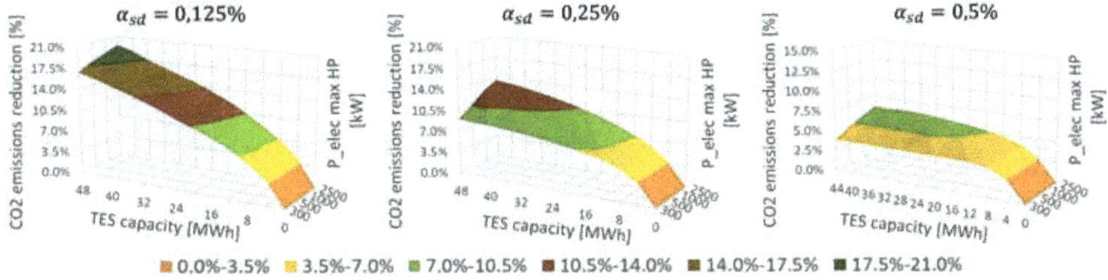

Figure 14. CO_2 emissions reduction according to the design of the heat pump and energy storage system.

Two phenomena happen when designing the energy system to use flexibility to reduce CO_2 emissions:

- The ability to store in the long-run: defined by the storage capacity and by the self-discharge coefficient. Indeed, with relatively poor insulation (α_{sd} = 0.5%), increasing the storage capacity beyond 8 MWh has no significant effect because of the importance of losses for long-term storage. However for a higher quality of insulation (α_{sd} = 0.125%), increasing the capacity until 48 MWh is always beneficial from an environmental point of view.
- The possibility to store a lot of energy during low-CO_2 periods: defined by the charging power of the storage and by the maximal power that can be consumed by the heat pump. In the case of a TES with a 48 MWh capacity and a 0.5% self-discharge coefficient, increasing the maximal power consumed by the heat pump from 300 kW to 2500 kW saves from 3.9% to 6.2%. Indeed, with higher electrical consumption, the heat pump can provide more low-CO_2 thermal power to the storage.

Although it seems reachable to have a strong impact on the CO_2 emissions of the district heating load with a big TES and a heat pump with high electrical power needs, this design choice leads to other problems. Indeed, choosing the kind of heat pumps means to increase the electrical power peaks and could report CO_2 emissions decreases from the heating side to increases at the electrical one. Over-sizing the heat pump should thus be carefully considered taking this effect in mind. Moreover, a 48 MWh water tank is expensive and takes a lot of space so that it is not an ideal solution. Nevertheless, it could be very interesting to deeply consider the level of insulation that can have an important impact.

Finally, using a building's envelope as storage should be investigated.

4.1.3. Study Case 2: Flexibility through Heating Loads Modulation

In this case, the CO_2 reduction is low (0.5%). Although the flexibility potential was previously estimated to 22%, it mainly relies on medium- and long-term flexibility. However, with the Building as Battery (BaB) concept, the flexibility addressed in our case can be defined as short term. Indeed, we can notice on the optimization results (Figure 15) that no energy is shifted for more than one day. High consumption peaks allow profiting from low-CO_2 rate periods but the energy cannot be stored in the long run. Indeed, with external wall insulation systems (EWIS), buildings envelopes form a relatively small storage capacity, while the CO_2 variability is more long-term in this case.

Moreover, over-heating the buildings leads to an increase in district energy consumption (0.9%), so that the environmental gain due to shifting the heating load is reduced. However, the mean operative temperature goes from 20.3 °C to 20.4 °C, i.e., an increase of 0.1 °C (0.4%). Therefore, the increase of the consumption induces a better thermal comfort while reducing CO_2 emissions. Many studies achieve a greater reduction by allowing over- and under-heating [47], i.e., by using both energy flexibility and sobriety, while we choose to focus on the impact of flexibility only.

Figure 15. Optimal flexibility to minimize CO_2 emissions by allowing over-heating.

With a maximal electrical consumption of heat pumps of 500 kW, this scenario can be compared to those with a single 500 kW heat pump. The reduction of CO_2 emissions obtained by buildings as storage is similar to results with a TES with a capacity between 100 kWh and 1 MWh, whatever the self-discharge coefficient.

Finally, the reference scenario is the result of an optimization problem by providing minimal energy to maintain thermal comfort. Thus, when over-heating is not compensated by the CO_2 diminution, the CO_2 emissions minimization corresponds to the energy minimization.

For this reason, we need to compare a simulation reference profile to the flexible scenario to see if the improvement is preserved with a standard controller. Besides, comparing OMEGAlpes results with simulation can lead to investigating optimal control robustness during application.

4.2. Simulation

Experimenting flexibility scenarios computed by OMEGAlpes directly on the real district is a complex task. Before performing these tests, we started simulation studies to validate our approach and/or identify the main issues towards implementation. Since we used TEASER as an intermediate for building OMEGAlpes models, Modelica models are also ready to simulate for each building. First, we can compare heating needs and operative temperature profiles (defined as a mean between air and radiant temperatures) for both modeling approaches. To do so, we used two specific control scenarios:

- **Constant temperature setp oint:** In this case, we want to achieve a constant operative temperature of 20 °C inside each building. With Modelica models, it consists in inserting an operative temperature sensor and regulating the injected heat power with a PI controller. In the case of OMEGAlpes models, the heat demand is computed to minimize the discrepancy between buildings operative temperatures and the 20 °C set point.
- **Flexibility scenario:** In the optimization case presented above, OMEGAlpes has reduced energy consumption and CO_2 emissions while preserving thermal comfort constraints. To reproduce

computed power shifts on Modelica models, we applied the operative temperatures computed for the flexibility scenario as a new setp oint profile.

In the first case of constant temperature set point, we obtained the results presented in Figure 16 (buildings mean operative temperatures and the sum of all buildings consumption).

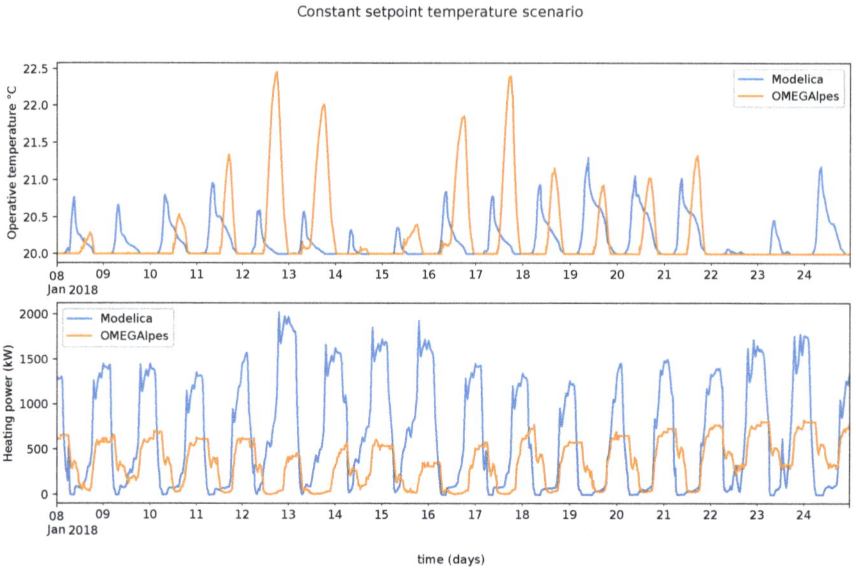

Constant setpoint temperature scenario

Figure 16. Comparison between Modelica and OMEGAlpes—Constant temperature set point scenario.

The first obvious difference stands in energy needs. Modelica models generated by TEASER have a more important consumption for the same comfort criteria and are therefore probably less insulated. Besides, temperature peaks are shifted between models. These shifts can be due to the differences in control strategies, and/or in computed internal gains despite using the same scenarios and weather files. Consequently, even with the same data sources, it appears hard to obtain identical dynamic behaviors between different modelers. Further effort must be invested to preserve global building characteristics during model translating (in our case, model simplification). Besides, this also suggests the use of a model calibration phase before implementing any model-based control strategy.

If we consider the application of the flexibility scenario in Figure 17, the consumption differences between both models are visually less important, except for higher spikes for Modelica models, but dynamics of operative temperature are sill very present (more inertia to go down for Modelica models).

We also compared performance results between constant temperature set point and computed flexibility temperatures on Modelica models only, to see if it also leads to improvement despite model discrepancies (see Figure 18).

Here, the dynamics induced by the flexibility scenario are very noticeable. As for OMEGAlpes results, we observe power shifts and spikes inducing heat storage in buildings envelopes. Unfortunately, performance is not preserved, since both energy consumption and CO_2 emissions are worsened (Table 2).

This implies that the flexibility command computed here is not robust to the model discrepancies we are facing. Therefore, the robustness of flexibility scenarios towards modeling uncertainties is certainly a key research topic before real-life integration if we do not want optimized scenarios to be counterproductive. This is also true during the early stage of design where low order models are used to investigate energy scenarios.

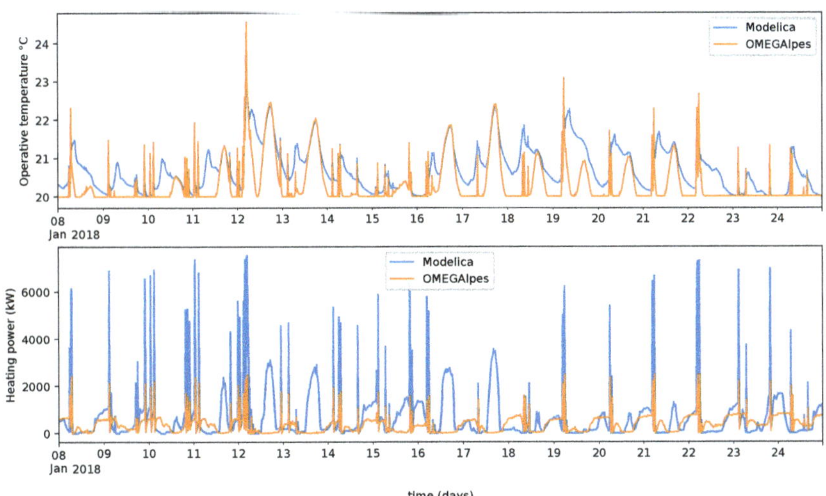

Figure 17. Comparison between Modelica and OMEGAlpes—flexibility scenario.

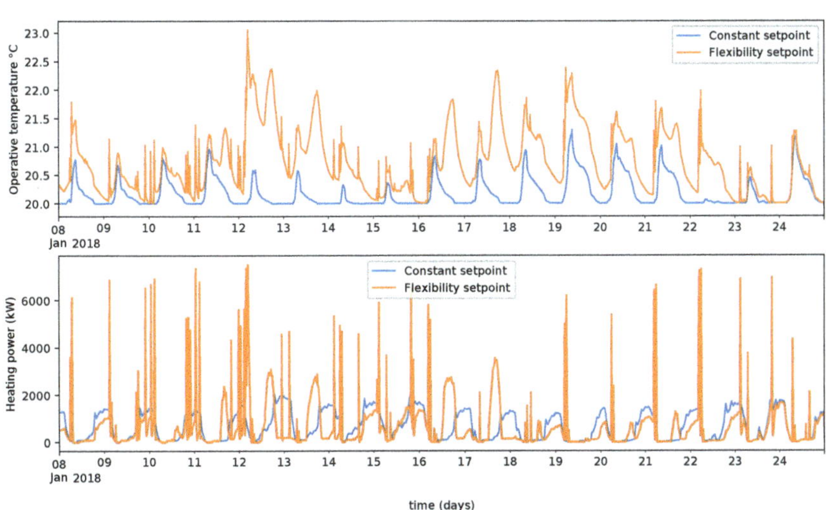

Figure 18. Modelica simulations—comparison between control scenarios.

Table 2. Performance indices with flexibility scenario to optimize CO_2 emissions against constant temperature setpoint.

	Energy Consumption	CO_2 Emissions
Improvement with OMEGAlpes model	−0.78%	0.41%
Improvement with Modelica model	−2.28%	−0.63%

5. Conclusions

Based on an ETL (Extract–Transform–Load) method, we have initiated a tool based on the heterogeneous available data of buildings at the district scale, which can generate the necessary data for optimization and simulation models. In particular, we have applied this method to the case of data available on a new residential district composed of 16 residential buildings. This work makes it possible to identify the important parameters for different modeling tools at the neighborhood scale, to extract them from available data, or to estimate them when they are not available.

We then carried out two flexibility studies, based on the OMEGAlpes tool, which requires modeling in MILP formulation. The first study analyzed the design of a heat pump (nominal power) and storage (capacity and self-discharge factor) to desynchronize the production of heat and use to heat buildings. It appears that a large investment is necessary to try to reach the maximum potential (estimated at 22%), which relies in particular on long-term flexibility (more than a week). The second study relied on thermal storage via the building envelope. This zero investment solution is, therefore, a potential alternative to the previous case. However, the results obtained below 1% show that the low storage capacity of these residential buildings does not allow addressing flexibility considering a CO_2 variability during several days.

Finally, the tool we developed for data processing at the neighborhood scale allowed us to easily set up a validation process. Thus, we have transmitted the flexibility results in the simulation model using the Modelica AixLib library. The results show a predicted performance degradation compared with optimization results. On very small gains (<1%) obtained by the upward flexibility (temperature > 20 °C), it even presents negative performance in energy and CO_2. This lack of robustness to modeling assumptions reinforces the idea that a tool to generate different levels of modeling based on available data will be indispensable for future studies related to robustness in optimization.

Author Contributions: Supervision, B.D., S.R., F.W. and Y.M.; Writing—original draft, C.P. and N.A.; and Writing—review and editing, C.P., N.A. and B.D.; C.P. and N.A. contributed equally to this work.

Funding: This work was partially supported by the ANR project ANR-15-IDEX-02.

Conflicts of Interest: The authors declare no conflict of interest.

References

1. Climate Action Tracker—EU. Available online: https://climateactiontracker.org/countries/eu/ (accessed on 30 July 2019).
2. International Energy Agency. Buildings. Available online: https://www.iea.org/topics/energyefficiency/buildings/ (accessed on 30 July 2019)
3. Kammen, D.M.; Sunter, D.A. City-integrated renewable energy for urban sustainability. *Science* **2016**, *352*, 922–928. [CrossRef] [PubMed]
4. Henning, D. MODEST—An energy-system optimisation model applicable to local utilities and countries. *Energy* **1997**, *22*, 1135–1150. [CrossRef]
5. Howells, M.; Rogner, H.; Strachan, N.; Heaps, C.; Huntington, H.; Kypreos, S.; Hughes, A.; Silveira, S.; DeCarolis, J.; Bazillian, M.; et al. OSeMOSYS: The Open Source Energy Modeling System: An introduction to its ethos, structure and development. *Energy Policy* **2011**, *39*, 5850–5870. [CrossRef]
6. Gardumi, F.; Shivakumar, A.; Morrison, R.; Taliotis, C.; Broad, O.; Beltramo, A.; Sridharan, V.; Howells, M.; Hörsch, J.; Niet, T.; et al. From the development of an open-source energy modelling tool to its application and the creation of communities of practice: The example of OSeMOSYS. *Energy Strategy Rev.* **2018**, *20*, 209–228. [CrossRef]
7. Messner, S.; Schrattenholzer, L. MESSAGE–MACRO: Linking an energy supply model with a macroeconomic module and solving it iteratively. *Energy* **2000**, *25*, 267–282. [CrossRef]
8. Loulou, R.; Labriet, M. ETSAP-TIAM: The TIMES integrated assessment model Part I: Model structure. *Comput. Manag. Sci.* **2008**, *5*, 7–40. [CrossRef]
9. Loulou, R. ETSAP-TIAM: The TIMES integrated assessment model. Part II: Mathematical formulation. *Comput. Manag. Sci.* **2008**, *5*, 41–66. [CrossRef]

10. Prospective Outlook on Long-Term Energy Systems. Available online: https://ec.europa.eu/jrc/en/poles (accessed on 30 July 2019).
11. Criqui, P.; Mima, S.; Viguier, L. Marginal abatement costs of CO_2 emission reductions, geographical flexibility and concrete ceilings: An assessment using the POLES model. *Energy Policy* **1999**, *27*, 585–601. [CrossRef]
12. Cochran, J.; Bird, L.; Heeter, J.; Arent, D. *Integrating Variable Renewable Energy in Electric Power Markets: Best Practices from International Experience*; (No. NREL/TP-6A20-53732); National Renewable Energy Lab. (NREL): Golden, CO, USA, 2012.
13. Taibi, E.; Nikolakakis, T.; Gutierrez, L.; Fernandez, C.; Kiviluoma, J.; Rissanen, S.; Lindroos, T.J. *Power System Flexibility for the Energy Transition: Part 1, Overview for Policy Makers*; International Renewable Energy Agency IRENA: Abu Dhabi, UAE, 2018.
14. Lund, P.D.; Lindgren, J.; Mikkola, J.; Salpakari, J. Review of energy system flexibility measures to enable high levels of variable renewable electricity. *Renew. Sustain. Energy Rev.* **2015**, *45*, 785–807. [CrossRef]
15. Mendes, G.; Loakimidis, C.; Ferrão, P. On the planning and analysis of Integrated Community Energy Systems: A review and survey of available tools. *Renew. Sustain. Energy Rev.* **2011**, *15*, 4836–4854. [CrossRef]
16. Keirstead, J.; Jennings, M.; Sivakumar, A. A review of urban energy system models: Approaches, challenges and opportunities. *Renew. Sustain. Energy Rev.* **2012**, *16*, 3847–3866. [CrossRef]
17. Allegrini, J.; Orehounig, K.; Mavromatidis, G.; Ruesch, F.; Dorer, V.; Evins, R. A review of modelling approaches and tools for the simulation of district-scale energy systems. *Renew. Sustain. Energy Rev.* **2015**, *52*, 1391–1404. [CrossRef]
18. HOMER—Hybrid Renewable and Distributed Generation System Design Software. Available online: https://www.homerenergy.com/ (accessed on 30 July 2019).
19. Simpkins, T.; Cutler, D.; Anderson, K.; Olis, D.; Elgqvist, E.; Callahan, M.; Walker, A. REopt: A platform for energy system integration and optimization. In Proceedings of the ASME 2014 8th International Conference on Energy Sustainability Collocated with the ASME 2014 12th International Conference on Fuel Cell Science, Engineering and Technology, Boston, MA, USA, 30 June–2 July 2014; pp. ES2014-6570, V002T03A006.
20. Artelys | Optimization Solutions—Artelys Crystal City. Available online: https://www.artelys.com/fr/crystal/city/ (accessed on 30 July 2019).
21. Bollinger, L.A.; Dorer, V. The Ehub Modeling Tool: A flexible software package for district energy system optimization. *Energy Procedia* **2017**, *122*, 541–546. [CrossRef]
22. Distributed Energy Resources—Customer Adoption Model (DER-CAM) | Building Microgrid. Available online: https://building-microgrid.lbl.gov/projects/der-cam (accessed on 30 July 2019).
23. Hilpert, S.; Kaldemeyer, C.; Krien, U.; Günther, S.; Wingenbach, C.; Plessmann, G. The Open Energy Modelling Framework (oemof)—A new approach to facilitate open science in energy system modelling. *Energy Strategy Rev.* **2018**, *22*, 16–25. [CrossRef]
24. Atabay, D. An open-source model for optimal design and operation of industrial energy systems. *Energy* **2017**, *121*, 803–821. [CrossRef]
25. Ringkjøb, H.K.; Haugan, P.M.; Solbrekke, I.M. A review of modelling tools for energy and electricity systems with large shares of variable renewables. *Renew. Sustain. Energy Rev.* **2018**, *96*, 440–459. [CrossRef]
26. Pipattanasomporn, M.; Kuzlu, M.; Rahman, S.; Teklu, Y. Load Profiles of Selected Major Household Appliances and Their Demand Response Opportunities. *IEEE Trans. Smart Grid* **2013**, *5*, 742–750. [CrossRef]
27. Laicane, I.; Blumberga, D.; Blumberga, A.; Rosa, M. Reducing Household Electricity Consumption through Demand Side Management: The Role of Home Appliance Scheduling and Peak Load Reduction. *Energy Procedia* **2015**, *72*, 222–229. [CrossRef]
28. Reynders, G.; Nuytten, T.; Saelens, D. Potential of structural thermal mass for demand-side management in dwellings. *Build. Environ.* **2013**, *64*, 187–199. [CrossRef]
29. Oliveira Panão, M.J.N.; Mateus, N.M.; Carrilho da Graça, G. Measured and modeled performance of internal mass as a thermal energy battery for energy flexible residential buildings. *Appl. Energy* **2019**, *239*, 252–267. [CrossRef]
30. Arteconi, A.; Polonara, F.; Arteconi, A.; Polonara, F. Assessing the Demand Side Management Potential and the Energy Flexibility of Heat Pumps in Buildings. *Energies* **2018**, *11*, 1846. [CrossRef]
31. Le Dréau, J.; Heiselberg, P. Energy flexibility of residential buildings using short term heat storage in the thermal mass. *Energy* **2016**, *111*, 991–1002. [CrossRef]

32. Pajot, C.; Morriet, L.; Hodencq, S.; Reinbold, V.; Delinchant, B.; Wurtz, F.; Maréchal, Y. Omegalpes: An Optimization Modeler as an Efficient Tool for Design and Operation for City Energy Stakeholders and Decision Makers. In Proceedings of the 2019 IBPSA Building Simulation International Conference, Rome, Italy, 2–4 September 2019.
33. Data MetropoleGrenoble—Saisissez vous des Données. Available online: http://data.metropolegrenoble.fr/ (accessed on 30 July 2019).
34. EU Buildings Database. Available online: https://ec.europa.eu/energy/en/eu-buildings-database (accessed on 30 July 2019).
35. PSS—ARCHI EU. Available online: http://www.pss-archi.eu/ (accessed on 30 July 2019).
36. Joint Website of the TABULA and EPISCOPE Projects. Available online: http://episcope.eu/welcome/ (accessed on 30 July 2019).
37. Kimball, R.; Caserta, J. *The Data Warehouse ETL Toolkit: Practical Techniques for Extracting, Cleaning, Conforming, and Delivering Data*; John Wiley & Sons, Wiley Publishing Inc.: Indianapolis, IN, USA, 2011.
38. Walter, E.; Kämpf, J.H. A verification of CitySim results using the BESTEST and monitored consumption values. In Proceedings of the 2nd Building Simulation Applications Conference, Bolzano, Italy, 4–6 February 2015; pp. 215–222.
39. Fonseca, J.A.; Nguyen, T.A.; Schlueter, A.; Marechal, F. City Energy Analyst (CEA): Integrated framework for analysis and optimization of building energy systems in neighborhoods and city districts. *Energy Build.* **2016**, *113*, 202–226. [CrossRef]
40. Remmen, P.; Lauster, M.; Mans, M.; Fuchs, M.; Osterhage, T.; Müller, D. TEASER: An open tool for urban energy modelling of building stocks. *J. Build. Perform. Simul.* **2018**, *11*, 84–98. [CrossRef]
41. Chen, Y.; Hong, T.; Piette, M. Automatic Generation and Simulation of Urban Building Energy Models Based on City Datasets for City-Scale Building Retrofit Analysis. *Appl. Energy* **2017**, *205*. [CrossRef]
42. Sola, A.; Corchero, C.; Salom, J.; Sanmarti, M. Simulation tools to build urban-scale energy models: A review. *Energies* **2018**, *11*, 3269. [CrossRef]
43. Lauster, M. AixLib.ThermalZones.ReducedOrder.RC.FourElements. Available online: https://build.openmodelica.org/Documentation/AixLib.ThermalZones.ReducedOrder.RC.FourElements.html (accessed on 30 July 2019).
44. Trujillo, J.; Luján-Mora, S. A UML Based Approach for Modeling ETL Processes in Data Warehouses. In Proceedings of the International Conference on Conceptual Modeling, Chicago, IL, USA, 13–16 October 2003.
45. Raccanello, J.; Rech, S.; Lazzaretto, A. Simplified dynamic modeling of single-tank thermal energy storage systems. *Energy* **2019**, *182*, 1154–1172. [CrossRef]
46. Pajot, C.; Delinchant, B.; Maréchal, Y.; Artiges, N. Building Reduced Model for MILP Optimization: Application to Demand Response of Residential Buildings. In Proceedings of the 2019 IBPSA Building Simulation international conference, Rome, Italy, 2–4 September 2019.
47. Kampelis, N.; Sifakis, N.; Kolokotsa, D.; Gobakis, K.; Kalaitzakis, K.; Isidori, D.; Cristalli, C. HVAC Optimization Genetic Algorithm for Industrial Near-Zero-Energy Building Demand Response. *Energies* **2019**, *12*, 2177. [CrossRef]

Article

Combined Optimal Design and Control of Hybrid Thermal-Electrical Distribution Grids Using Co-Simulation

Edmund Widl [1,*], Benedikt Leitner [1], Daniele Basciotti [1], Sawsan Henein [1], Tarik Ferhatbegovic [1] and René Hofmann [1,2]

1 Center for Energy, Austrian Institute of Technology, 1210 Vienna, Austria; benedikt.leitner@ait.ac.at (B.L.); daniele.basciotti@ait.ac.at (D.B.); sawsan.henein@ait.ac.at (S.H.); tarik.ferhatbegovic@ait.ac.at (T.F.); rene.hofmann@ait.ac.at (R.H.)

2 Institute for Energy Systems and Thermodynamics, Vienna University of Technology, TU Wien, 1060 Vienna, Austria

* Correspondence: edmund.widl@ait.ac.at

Received: 28 February 2020; Accepted: 8 April 2020; Published: 15 April 2020

Abstract: Innovations in today's energy grids are mainly driven by the need to reduce carbon emissions and the necessary integration of decentralized renewable energy sources. In this context, a transition towards hybrid distribution systems, which effectively couple thermal and electrical networks, promises to exploit hitherto unused synergies for increasing efficiency and flexibility. However, this transition poses practical challenges, starting already in the design phase where established design optimization approaches struggle to capture the technical details of control and operation of such systems. This work addresses these obstacles by introducing a design approach that enables the analysis and optimization of hybrid thermal-electrical distribution systems with explicit consideration of control. Based on a set of key prerequisites and modeling requirements, co-simulation is identified as the most appropriate method to facilitate the detailed analysis of such systems. Furthermore, a methodology is presented that links the design process with the implementation of different operational strategies. The approach is then successfully applied to two real-world applications, proving its suitability for design optimization under realistic conditions. This provides a significant extension of established tools for the design optimization of multi-energy systems.

Keywords: design optimization; control and operation; multi-carrier energy systems; co-simulation

1. Introduction

The joint design and integrated operation of electrical distribution grids and district heating systems promises to exploit hitherto unused synergies for increasing efficiency and flexibility. The envisaged goal is to achieve an increase of the hosting capability of electrical distribution networks for renewable energy sources (RES), while simultaneously reducing greenhouse gas emissions and primary energy use of district heating systems. For instance, in electrical distribution grids, the integration of photovoltaic (PV) and wind generation into the existing infrastructure has severe consequences on the power quality. At the same time, district heating networks struggle to replace carbon intensive heat plants with economically feasible combined heat and power plants and other renewable heat sources. In case network planning and operation are done appropriately, the diverse storage technologies, the different time constants and the diverse constraints regarding demand and generation can complement each other.

However, there is a lack of tools and methods for designing such integrated energy systems, especially in view of a detailed validation of proposed control and operation schemes. The work presented here introduces a design approach that addresses the technical challenges of both domains at the same time, including their dynamic interaction at network level as well as local and high-level control (see Figure 1). This enables the exploitation of synergies in the operation of hybrid thermal-electrical distribution system in an optimal way. The presented approach relies on co-simulation, which enables the coupling of domain-specific simulators and multi-purpose tools in a way that allows to combine multi-physics simulations and optimization procedures. This approach is motivated by the criteria established in Section 2, and backed up by reviewing the available tools and methods according to the state-of-the-art for simulation and optimization (Section 3). A description of the modeling approach and the utilized simulation tools is presented in Section 4. Based on this, a novel methodology is presented in Section 5 that links design constraints to suitable operational strategies and optimization methods. Finally, the applicability of the presented approach is demonstrated in two real-world applications in Section 6.

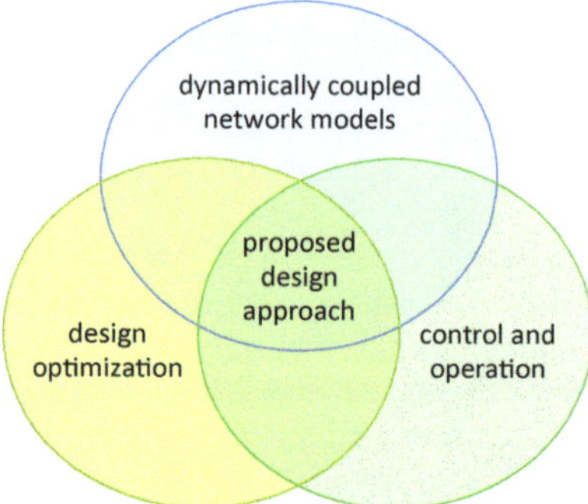

Figure 1. Scope of the design approach proposed in this work.

Scope and Main Contributions

This paper aims at developing an integrated optimal design and control framework for coupled district heating and electrical distribution networks, extending the scope of traditional design tools for multi-energy systems. The simulation and optimization framework is illustrated for designing storage and thermal-electric appliances in two case studies, i.e., an industrial and a rural area. Note that the focus of this work lies on the methodological contribution rather than on the case study results. Compared to existing studies, this work explicitly includes detailed models for the district heating network, the electric distribution grid as well as low- and high-level control implementations already in the design stage. Thus, the method considers the impact of different levels of control and operation on the optimal system design. The coupling of heterogeneous modeling paradigms and tools is established via a co-simulation approach and the design optimization relies on the use of meta-heuristics.

This work addresses experts with backgrounds in district heating, electric distribution, energy storage deployment, control theory, design optimization and (co-)simulation alike. The multi-disciplinary nature of the proposed design approach (see also Figure 1) requires a comprehensive introduction and presentation of the work to make it understandable and useful for readers and experts coming from these different fields.

In summary, this paper contributes to the research field by presenting a simulation-based design optimization approach that: (i) is based on a fully dynamic thermal-hydraulic district heating and electrical distribution network model; and (ii) explicitly includes closed-loop control implementations and, thus, leads to the optimal design being dependent on operational aspects. The method is used in two case studies that exhibit different control complexities, i.e., model predictive and rule-based, and design spaces, i.e., a finite set of allowed solutions and an infinite set of possible designs.

2. Prerequisites for the Design Optimization of Hybrid Distribution Systems

2.1. Determining Factors of Hybrid Distribution Systems

Hybrid networks are realized through the physical interconnection and joint operation of electrical and thermal distribution grids. From a technical perspective, this is accomplished with the help of *coupling points*, i.e., devices which directly or indirectly enable the exchange of energy across carrier domains. For the proposed design approach, a (preliminary) *technical system layout* for both the electrical network and the thermal network topology is required from the domain experts, which already includes the type and position of the coupling point(s). The considered *degrees of freedom* in this work are typically related to the sizing of components (e.g., storage capacities or power ratings) or controller set-points (e.g., gains or thresholds). A specific technical system layout together with a specific set of numerical values for theses degrees of freedom is referred to as *system configuration* in this work.

The proposed design optimization approach explicitly considers operational and control aspects at different levels. In general, there are control systems at the process level (e.g., for heat pumps or transformers) that are designed to ensure that objectives are achieved locally (e.g., valve positions and tap changer). However, the complexity of control schemes increases drastically when individual processes are combined to larger systems and new (common) control/optimization targets are defined. In such a case, a higher-level control instance—referred to as *operational strategy* in this work—is required that governs the local processes in compliance with relevant system constraints (compare with Figure 2).

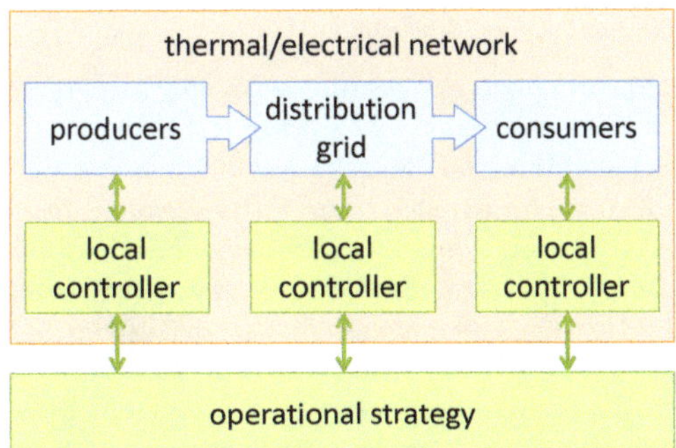

Figure 2. Overview of local control (process level) versus operational strategy (system level).

2.2. Key Prerequisites and Modeling Requirements

This work focuses on the optimization of technical design aspects, assessing the impact of system configurations and operational strategies on the system's performance. This requires not only the quantification of performance indicators on the system level, but also validation down to the component level, in order to avoid infeasible results. Hence, any valid design approach for hybrid thermal-electrical distribution networks has to fulfill the *key prerequisites* (KP) presented in Table 1.

In general, the most viable approaches in this context are simulation-based methods, as they allow the qualitative and quantitative examination of new concepts in a comparably fast and inexpensive way before deployment. However, the accurate and detailed modeling of all involved networks including their coupling points and control is of utmost importance. When deployed in a joint simulation, the (sub-)models representing the individual domains have to comply with the *modeling requirements* (MR) presented in Table 1.

It should be noted that some of the criteria defined above do not necessarily apply for design optimization approaches focusing on socioeconomic goals, especially regarding the required level of detail (spatial and temporal).

Table 1. Key prerequisites and modeling requirements for the assessment of hybrid networks.

	Key Prerequisites
KP1	A method for analyzing coupled thermal and electrical distribution systems on a network level is required, which allows a technical assessment with reasonably high spatial and temporal resolution.
KP2	Given a system layout with certain degrees of freedom and a design criterion represented by a (scalar) objective function, a suitable method is needed to optimize this layout by minimizing this objective function.
KP3	The tools and methods used for analysis and optimization must enable domain experts to actively participate in the design process.
	Modeling Requirements
MR1	When combined, the (sub-)models need to capture the dynamic interactions between the domains, i.e., electric, thermal and control, on the system level. This is the prerequisite for enabling the design of new coupling concepts and operational strategies for hybrid distribution systems.
MR2	At the same time, they have to enable the detailed study of the impact on the individual domains, from the network down to the component level. This is required to check the feasibility of any coupling concept and/or operational strategy from the technical perspective of the respective domains.
MR3	In addition, the proper representation of local controls at process level and overall operational strategies is required. Their proper interaction with the models of the physical processes is fundamental for a complete system representation and can come in various forms, e.g., from simple PID to model predictive control.

3. State of the Art Regarding the Simulation and Optimization of Hybrid Distribution Systems

3.1. Simulation of Coupled Heat and Power Networks

Previous work on multi-carrier energy systems was mainly focused on determining an optimal mix of energy sources as well as conversion and storage technologies. Comprehensive overviews were given by Mancarella et al. [1] and van Beuzekom et al. [2], who independently concluded that existing approaches are not suited for detailed technical assessments that are required for network infrastructure planning and operation. Consequently, they argued that there is a need for new tools and methods in this regard. One of the main reasons is that established tools for system design, such as EnergyPLAN [3] or HOMER [4], do not offer models that are detailed enough to evaluate the effects of local controls on the process level, thus disagreeing with KP1, MR2 and MR3. Unfortunately, simulation tools that focus on the technical evaluation of energy systems on the process level are by themselves also not suited in this context [5], as they are typically concerned with just a single energy-related domain. Anything outside their direct focus is usually taken into account only implicitly or using simplified models, thus disagreeing with KP1, MR1, MR2 and MR3.

A potential solution is provided by multi-domain modeling languages. For instance, there exist several libraries for the Modelica [6] language that target energy-related domains, such as power systems [7,8] or buildings [9], which in combination allow the modeling of hybrid thermal-electrical distribution systems (and thus satisfying KP1 and MR1). It has also been demonstrated how such models can be utilized for standard optimization approaches (satisfying KP2), see for instance [10]. However, domain experts (thermal, electrical, controls) are often not trained to use such tools, which is complicating compliance with KP3. Furthermore, since Modelica focuses on modeling physical systems, the implementation of controls and operational strategies with a high algorithmic effort, e.g., model predictive control, can become difficult, thus complicating compliance with MR3. Most of the above arguments are also valid for similar languages and tools, e.g., MATLAB/Simulink/Simscape [11]. Hence, in the context of simulating hybrid thermal-electrical distribution networks, the usability of approaches relying exclusively on multi-domain modeling languages is in practice (still) limited, or at least the associated effort is (still) considerably high.

An alternative solution is offered by co-simulation approaches, which overcome these limitations by enabling the coupling of domain-specific simulators and multi-purpose tools (thus satisfying MR2 and MR3). This allows domain experts to use the most appropriate tools according to the state of the art for their respective domain, including advanced optimal control schemes. On the one hand, this guarantees an adequate and precise representation of the individual domains (thus satisfying KP1 and MR1) [12]. On the other hand, it facilitates the participation of and interaction between experts from different domains, whose expertise is often closely linked to specific tools (thus satisfying KP3). This advantage has led to various research activities in many energy-related domains, e.g., buildings [13,14], power systems [15–18] and hybrid distribution networks [12,19].

Given these considerations, co-simulation has been chosen to analyze systems according to the criteria defined in Section 2 for the work presented here. Even though other approaches—and especially approaches based on multi-domain modeling languages—are expected to become mature and flexible enough in the future, the advantages of coupling domain-specific simulators and multi-purpose tools in a co-simulation still prevail.

Within this context, established tools for the design and optimization of multi-energy systems can be considered as important guides for an overall design process, as their results should be used as starting point for a detailed technical evaluation.

3.2. Design Optimization of Coupled Heat and Power Networks

Available literature on design optimization of hybrid thermal-electric networks is mainly based on the energy hub concept introduced by Geidl et al. [20]. Related work focuses on determining optimal design of energy hubs such as selection and sizing of coupling units and storages [21]. However, applied models rely on many simplifications and are not able to cover technical aspects relevant in power networks (e.g., voltages and reactive power), in district heating (e.g., temperature propagation and pressures), and also in individual components (e.g., temperature stratification in thermal storages). Thus, such an approach contradicts the identified modeling requirements MR2 as well as MR3.

One possible solution, targeted in this paper, is to utilize detailed (co-)simulation setups for design optimization. Even though simulations are very well suited to characterize a given system design using a system performance measure, their application in the context of design optimization is more challenging. From a mathematical programming point-of-view, the simulation-based design leads to objective functions that must, in general, be considered non-linear, multi-modal and discontinuous [22]. To make matters more complicated, the evaluation of the objective function is computationally expensive (minutes to hours or even days per evaluation), depending on the complexity of the simulation model. With the number of possible design variables being high and the range of corresponding input parameter values being huge or even infinite, it becomes infeasible to perform this search by hand or by brute-force. Thus, this problem class requires to either reduce the allowed solution space and/or to use efficient black-box optimization algorithms, where finding a global optimum within finite time is not guaranteed [23].

Nevertheless, simulation-based design optimization is a frequently used technique as it allows the use of detailed models. The possibility to use high-fidelity models for optimization is especially relevant when targeting systems-of-systems, such as hybrid thermal-electric networks in this work, and, thus, satisfies KP2. In energy-related research, applications to building design are most frequent. The dynamic simulation tool IDA ICE together with a multi-objective optimization was used to find cost-optimal energy performance renovation measures for educational buildings by Niemela et al. [24]. Delgarm et al. [25] used the building energy simulation program EnergyPlus and a multi-objective particle swarm optimization to find the building specifications that minimize annual energy consumption. Many more similar studies exist and Nguyen et al. [26] provided an extensive summary of building related simulation-based optimization methods. Recent work uses simulation-based optimization in the context of district heating system design. Wang et al. [27] optimized the hydraulic design of variable-speed pumps in multi-source district heating networks using static hydraulic simulation and a genetic optimizer. Van der Heijde et al. [28] tried to find the cost optimal location and size of thermal storage tanks in district heating networks using dynamic thermal-hydraulic simulation combined with optimal control and a genetic multi-objective optimization algorithm.

3.3. Summary and Conclusion Regarding the Relevant State of the Art

Based on the review of the relevant state of the art above, it follows that established optimization models do not provide the level of detail required for assessing hybrid thermal-electrical distribution grids, especially in view of closed-loop control and operation. In practice, currently only the co-simulation of domain-specific technical models can provide the desired degree of accuracy. However, even though these technical models are commonly used by domain experts for detailed assessments, they are typically not designed nor intended to be used for optimization. The literature gives examples of how co-simulation and this class of technical models can be used for optimization, but there exists—to best knowledge of the authors—no general methodology to support this process.

In this context, this work is the first to introduce simulation-based optimization for coupled district heating and electric network simulation combined with closed-loop control. A methodology to assist the co-simulation-based design of coupled heat and power networks is presented by providing conceptual guidance and a proof-of-concept implementation to the above mentioned problem setting. The focus lies on the utilized technical models (see Section 4) and their integration into suitable optimization approaches (see Section 5), not on the formulation of specific optimization algorithms.

Table 2 summarizes this situation in terms of the KPs and MRs identified above (+, full compliance; ∘, compliance with effort; −, insufficient compliance).

Table 2. Comparison of modeling and simulation approaches in view of key prerequisites and modeling requirements.

	KP1	KP2	KP3	MR1	MR2	MR3
established design optimization tools	−	+	∘	+	−	−
domain-specific tools	−	∘	+	−	−	−
multi-domain languages	+	∘	∘	+	+	−
proposed approach (based on co-simulation)	+	∘	+	+	+	+

4. Simulation Approach for Hybrid Networks

This section presents the approach used in this work for the detailed simulation of coupled heat and power networks including control. The overall model is highly complex, exhibits non-linear behavior and has no closed algebraic formulation. On the one hand, this rules out the utilization of the most commonly used optimization approaches (e.g., LP or MILP). On the other hand, this additional complexity is unavoidable for analyzing the technical details of control and operation in such networks. In view of the general optimization methodology presented in Section 5, the presented approach can be regarded as a representative example of co-simulation approaches used for technical assessments. As such, it highlights the differences between the class of simulation models targeted by this work and the class of models typically used for optimization.

4.1. Co-Simulation Environment

A co-simulation approach enables the coupling of the different modeling paradigms, i.e., a transient thermal-hydraulic model, a quasi-static power flow model and time-discrete advanced control models. Thus, the influence of time-discrete *advanced control systems*, e.g., using rule-based control or model predictive control (MPC), on the dynamic *physical system*, i.e., the electric and the district heating (DH) network including supplies, coupling units and consumers, can be studied. This enables the assessment of hybrid thermal-electrical distribution grids with appropriate spatial and temporal resolution for relevant use cases [12].

The modeling activities and environments used in this work are split into the control model and the physical system model, with the latter only including low-level control, e.g., PID controllers. The assessment method is based on modeling tools according to the state of the art for each domain that are presented in more detail in the following sections.

Within the context of this work, the FUMOLA environment [29] has been used as co-simulation environment. FUMOLA is specifically designed to support the features offered by the Functional Mock-up Interface (FMI) specification [30], which defines a standardized application programming interface (API) and model description for both co-simulation and model exchange. FMI was selected as it is a mature, non-proprietary specification, developed by both academia and industry. The FMI standard enables the exchange and extension of tools and methods for the different domains and, thus, makes the approach highly versatile and extensible, especially in selecting the most appropriate method for advanced control system implementation.

4.2. District Heating Network Model

Thermal networks (including producers, network, thermal storages and consumers) are modeled in Modelica/Dymola [6,31] with the help of the DisHeatLib library [32] that is built upon the IBPSA library [33]. These open-source libraries include models for the most relevant components of district heating networks, considering bi-directional mass flows, heat transport delays, detailed substation and storage models and other thermo-hydraulic aspects that are highly relevant in heat networks. In addition, it provides models of local controllers and interfaces to electric networks. In summary, this modeling approach captures transient thermal and quasi-static hydraulic network behavior. The most relevant models are shortly presented in the following.

All DH pipes are modeled using a plug flow approach. The outlet temperature T_{out} and, thus, the heat loss of a fluid parcel passing a pipe is described as:

$$T_{\text{out}} = T_{\text{g}} + \left(T_{\text{in}} - T_{\text{g}}\right) e^{-\frac{\tau}{R \cdot C}} \tag{1}$$

It depends only on the initial temperature T_{in}, residence time τ, undisturbed ground temperature T_{g} calculated using the Kusuda equation [34], thermal resistance of the pipe R and heat capacity of the water in the pipe C. The pressure drop mass flow correlation along the pipe is given by

$$\dot{m} = \text{sgn}(\Delta p) k \sqrt{|\Delta p|} \tag{2}$$

where k is the constant flow coefficient calculated for nominal conditions using the Colebrook equation for turbulent flow in rough pipes [35]. Details about implementation and experimental validation can be found in [36]. Heat exchangers in the DH substations are modeled with a variable effectiveness using a number of transfer units approach [37]. Valves are modeled using the above pressure drop and flow rate correlation with the flow coefficient k depending on the opening control signal.

Thermal energy storages are modeled using a vertically discretized multi-node approach to account for stratification and buoyancy [38]. Heat pumps are modeled using a Carnot-efficiency-based approach, electric heaters use a constant efficiency and gas and biomass boilers as well as combined heat and powers (CHPs) use heat generation dependent efficiencies. The main district heating supply unit is modeled as an ideal heat and differential pressure source and with no limits on maximum/minimum power or ramp rate. A fixed supply temperature is assumed for all generators.

A full list of model formulations and implementations can be found in the open-source Modelica libraries DisHeatLib and IBPSA.

4.3. Electrical Distribution Grid Model

Electrical distribution grids (including producers and consumers) are modeled with DIgSILENT PowerFactory [39], an engineering tool targeting primarily professional users. The quasi-static assessment of the electrical networks covers the most important features, such as voltage fluctuations and time-varying loads and generation. DIgSILENT PowerFactory's simulation interface has been extended to enable a series of consecutive power flow calculations in a co-simulation [40].

The power flow equations for node a in an N-node system are given in complex form:

$$\frac{S_a^*}{V_a^*} = \sum_{b=1}^{N} Y_{ab} V_b \tag{3}$$

where S_a^* and V_a^* denote the complex conjugate apparent power and voltage at node a, respectively, Y_{ab} denotes the bus admittance matrix and V_b denotes the complex voltage at node b. This results to $2n$ equations for the $4n$ unknowns, i.e., voltage, active/reactive power and voltage angle.

Coupling units are modeled as PQ buses, where active and reactive power is known, using the average power consumption/generation from the dynamic DH network model over the synchronization interval Δt_{qs} as input. Transformer units connect the low-voltage electrical networks to an external grid, i.e., modeled as a slack bus that determines the voltage and phase at the connection point.

4.4. Operation and Control Models

General-purpose tools such as MATLAB or Python can be integrated into the co-simulation with the help of FMI-compliant interfaces [41,42]. After each synchronization interval Δt_{ctrl}, they are called with the latest simulation outputs (measurement data), based on which they calculate and return new control setpoints that are then fed back to the physical models (feedback loop). This facilitates the implementation of a potentially large range of different types of algorithmic approaches for system-level controllers, including rule-based or model-predictive control schemes (see below). Local controllers on the process level are implemented within their respective subsystem models.

4.4.1. Rule-Based Operational Strategies

Rule-based control generally relies on a list of i deterministic rules formulated as logical and/or algebraic expressions, e.g., a hysteresis controller that issues on/off signals. These rules constitute the knowledge base to determine a control action $\boldsymbol{\psi}_i$ for a set of inputs (measurement data) corresponding to a certain system state $\boldsymbol{\phi}_i$.

$$(\boldsymbol{\phi}_1 \rightarrow \boldsymbol{\psi}_1) \wedge \ldots \wedge (\boldsymbol{\phi}_i \rightarrow \boldsymbol{\psi}_i) \tag{4}$$

Constructing this list of rules relies on expert knowledge. The reasoning depends on the specified operational goals and is often highly case-specific.

4.4.2. Model-Based Operational Strategies

Model-based control algorithms, in comparison, rely on knowledge about the dynamic system behavior $\dot{\mathbf{x}}(t)$ and the impact of control actions $\mathbf{u} \in \mathbf{U}$ to govern the overall system along an optimal trajectory. The model can be used within an optimal control scheme to determine the control action that satisfies all system dynamics and yields an optimal performance metric J:

$$\begin{aligned} \min_{\mathbf{u} \in \mathbf{U}} J &= \int_{t_s}^{t_f} L(\mathbf{x}(t), \mathbf{u}(t)) dt \\ s.t.\ \ \dot{\mathbf{x}}(t) &= f(\mathbf{x}(t), \mathbf{u}(t)) \end{aligned} \tag{5}$$

where L is a cost function, t_s is the start and t_f is the final time of the control horizon. The continuous time optimal control problem is often transformed into a time discrete version where the dynamics are represented by a (linear) state-space system. At runtime, feedback from the system (measurement data) and predictions of disturbances can be used to provide safe operation and optimal performance with respect to the operational goals. The use of model-based optimal control schemes is often labor intense, especially if model identification is not automated, requires a suitable optimization algorithm and solving the mathematical programming problem might be time consuming.

5. Design Optimization and Control of Hybrid Networks

This section describes the proposed design optimization framework for hybrid thermal-electrical distribution networks, focusing on the *technical assessment* of such systems for the purpose of network planning and operation. It is based on a simulation-based optimization approach that utilizes the detailed coupled heat and power network simulation including closed-loop control, presented in the previous section. The general resulting design optimization problem is given by

$$\mathbf{z}^* = \underset{\mathbf{z}\in\mathbf{\Omega}}{\operatorname{argmin}}\ c\left(g(\mathbf{z})\right) \tag{6}$$

where the function $g(\mathbf{z})$ involves one (co-)simulation run for a specific system configuration $\mathbf{z}$ and $\mathbf{\Omega}$ describes the solution space, i.e., the set of all possible and allowed system configurations. The goal of the design optimization process is to find the system configuration $\mathbf{z}^*$ that minimizes the objective function. Due to the high computational burden involved in executing one call of $g(\cdot)$, it is important to a priori reduce the number of possible system configurations. Hence, the proposed design approach assumes that most basic design decisions have already been reached using either expert know-how and/or established design optimization tools, e.g., the choice of conversion and storage technologies has been made employing mixed-integer linear programming techniques.

The *objective function* $c(\cdot)$ relates results from one simulation run $g(\mathbf{z})$ for a specific system configuration $\mathbf{z}$ to the optimization criterion, such as costs or technical key-performance indicators. Thus, the objective function maps certain technical and/or economical aspects of the overall system to a numerical (scalar) value. In the case of multi-carrier energy systems, objective functions typically relate aspects of the overall system that are traditionally treated by separate engineering domains, e.g, total energy imports for both heat and power. Furthermore, objective functions may evaluate effects that result from dynamic interactions between the subsystems, especially synergies among production, consumption and storage and their impact on network operation.

5.1. Influence of Operational Strategies on Optimal Design

The objective function of a given system configuration is highly dependent on the performance of the respective operational strategy and implemented control. Hence, to yield a small value for the objective function, the *operational goals* (e.g., the use of local PV generation for heat pump operation) should be in-line with the design *optimization targets* (e.g., sizing of heat pump to increase PV self-consumption). To this end, the operational goals and design optimization targets need to be translated into an appropriate control implementation.

In this work, two categories of control schemes, i.e., rule-based and model-based, are considered in more detail in Section 4.4. From the point of view of design optimization, these two categories of operational strategies serve very different purposes:

- The evaluation of a system configuration with the help of a rule-based operational strategy provides by itself little or no information about how this specific configuration could be improved. Improvements could be potentially achieved by changing either the design or the operational strategy.
- In contrast, the evaluation of a system configuration with the help of a model-based optimal operational strategy yields a measure for the best possible performance for this specific system configuration.

This difference comes from the fact that model-based optimal operational strategies are—by design—able to guide the system evolution in accordance to the defined optimization targets. In contrast, a rule-based approach is always heuristic, such that there is a priori no such guarantee.

In this case, both a different set of rules or a different system configuration could yield a performance improvement with respect to the objective function.

5.2. Design Optimization Approaches

The design process needs to adapt to the specific circumstances of any given project, especially constraints regarding available design options. For instance, there may be economical or legal restrictions or technical constraints (especially due to already existing infrastructure). In practice, this has a strong influence on the number of possible system configurations (see, for instance, the applications in Section 6). Hence, the following two complementary approaches are introduced, taking into account the size of the solution space and the implemented control scheme:

- *Optimal Control Scan* (OCS): In case only a very limited number of possible system configurations needs to be considered (e.g., due to specific design constraints), the evaluation of all these options with the help of an optimal control strategy determines the best possible design candidate.
- *Heuristic Parameter Scan* (HPS): In case the number of possible system configurations is large and the evaluation of all possible options is unfeasible due to the associated computational load, a metaheuristic optimization algorithm on top of a rule-based heuristic operational strategy can be utilized to determine the best possible design candidate.

In considerations of the above, the choice between OCS and HPS represents a trade-off among implementation effort, computational complexity and usability of the operational strategies for design optimization. Figure 3 visualizes this trade-off in terms of the size of the solution space, the computational complexity of the utilized operational strategy and the implementation effort for the associated sub-tasks (i.e., controller development and candidate selection). A combination of OCS and HPS, i.e., metaheuristic design optimization on top of an optimal control scheme, for a large number of possible system configurations, although preferable and theoretically possible, is not considered due to the high computational effort involved.

Figure 4 visualizes the different optimization approaches of OCS and HPS. In OCS, the optimizer is an integral part of the operational strategy implementation, guiding the system evolution towards optimal operational performance. In HPS, the optimizer is separated from the co-simulation and design optimization is achieved by repeatedly executing the co-simulation with different parameters. In this context, another benefit of using co-simulation becomes apparent, because the choice between both approaches has virtually no impact on the modeling of the thermal and electrical domain, as long as the operational strategy is implemented as an individual and exchangeable component in the co-simulation.

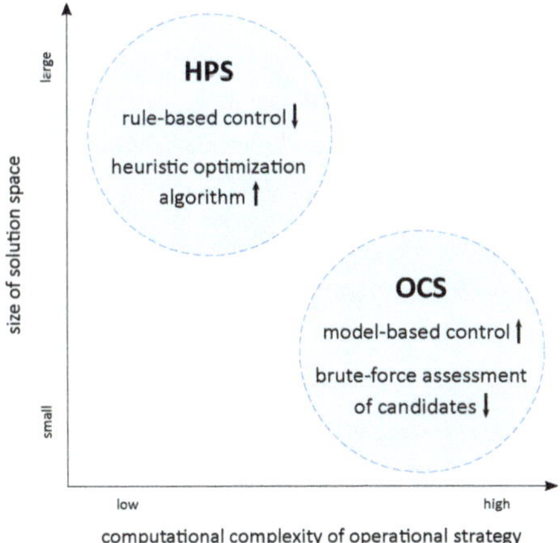

Figure 3. Comparison of the Optimal Control Scan (OCS) and the Heuristic Parameter Scan (HPS), indicating higher (↑) and lower (↓) implementation effort for specific sub-tasks.

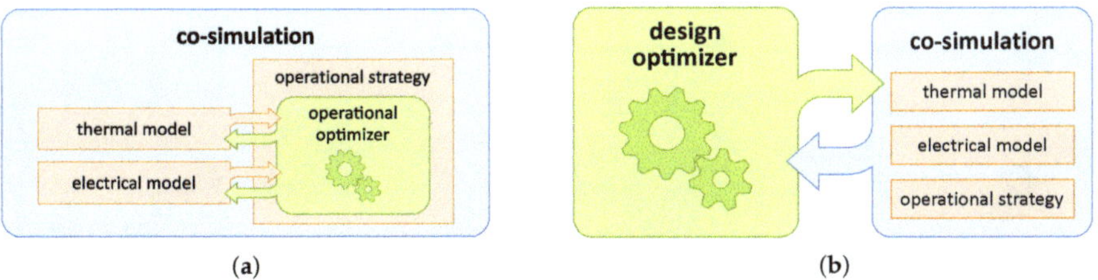

Figure 4. Schematic view of the relation between co-simulation and optimizer for: (**a**) the Optimal Control Scan (OCS); and (**b**) the Heuristic Parameter Scan (HPS).

5.3. Optimization Algorithms, Decision Variables and Objective Functions

The choice between OCS and HPS is first and foremost conceptual, providing a guideline for implementing the actual optimization. For both cases, a large variety of optimization algorithms exists that can be applied, thanks to the flexibility of co-simulation approaches. In the case of HPS, any metaheuristic population-based optimization algorithm can be applied that can take the results from individual co-simulation runs as black-box input, such as PSO [43], Differential Evolution [44] or PSwarm [45]. In the case of OCS, the co-simulation approach enables for instance the integration of existing toolboxes for model-predictive control, LP or MILP for MATLAB or Python. For example implementations—without loss of generality—refer to Section 6.

Using co-simulation of domain-specific models as basis for system assessment allows including more detailed technical information for decision variables compared to traditional optimization approaches. However, the choice between OCS and HPS does have practical implications for the selection of decision variables and objective functions. In the case of HPS, the decision variables used by the optimal control instance have to be based on the measurement data of the current and previous simulation time steps. For HPS, in contrast, the optimizer has not only access to all measurement data but also the full set of results of each completed co-simulation run. This means that overall key performance indicators (e.g., total energy saving or yearly local self-consumption) can be included in the objective function. This distinction is key for understanding the conceptual and qualitative difference regarding the optimality of results in OCS and HPS. Nevertheless, from a quantitative and practical point of view, both approaches yield (near) optimal results.

6. Proof-of-Concept Applications

The applicability and usefulness of the proposed design approach is demonstrated with the help of two real-world applications. Both applications focus on the design optimization of hybrid thermal-electrical grids from a technical perspective, aiming at exploiting synergies between the networks by mutual support during operation. Hence, no monetary optimizations are applied, i.e., neither investment nor operational costs are considered.

Both application examples aim at local consumption of excess PV generation using heat pumps or electric boilers. However, the proposed design approach is not limited to this kind of applications and could also be applied to applications with a focus on short-term and/or long-term storage, peak shaving, or others.

6.1. OCS Example Application: Suburban Industrial Area

6.1.1. Technical System Layout

The system is located in a suburban industrial park area, comprising multiple office buildings and industrial facilities at four adjacent sites, with a total yearly electrical demand of 1 GWh_{el} and a total yearly thermal demand of 2.3 GWh_{th}. Figure 5 presents a schematic of the technical system layout.

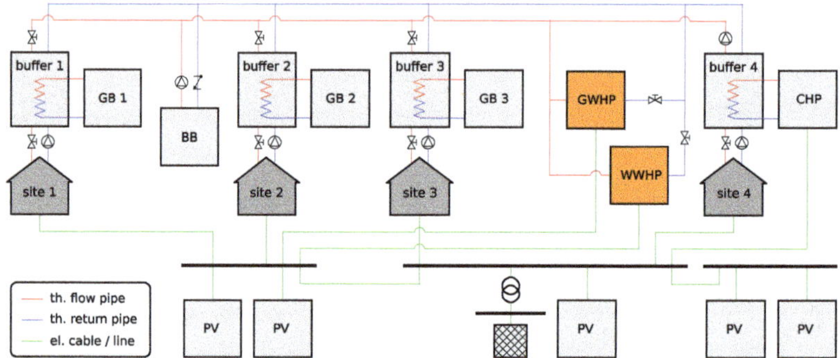

Figure 5. Schematic overview of the hybrid system layout for the suburban industrial area (GB, gas boiler; BB, biomass boiler; GWHP, ground water heat pump; WWHP, waste water heat pump; PV, photovoltaic module; CHP, combined heat and power plant). The components targeted by the design optimization process are highlighted in orange.

The on-site low voltage network consists of 0.6 km of cables and 0.6 km of overhead lines, and has five medium-size PV systems with a total installed capacity of 272 kWp_{el} connected to it. The heat network connects the thermal generators and demand sites, which makes it possible to share heat between them. Three of the sites use previously installed gas boilers (GBs) with a total nominal capacity of 2.44 MW_{th}, whereas the fourth site uses a CHP plant with a nominal capacity of 950 kW_{th}, all four connected to thermal buffer tanks. Moreover, a biomass boiler (BB) with 950 kW_{th} nominal capacity is feeding the heat network.

The main task in this application was to add and size a ground water heat pump (GWHP) and a waste water heat pump (WWHP), representing the *degrees of freedom* in the design process. These heat pumps and the CHP are the system's *coupling points* between the networks.

6.1.2. Operational Strategy

The foremost goal of the operational strategy is to maintain an operational temperature between 80 and 95 °C in all thermal buffers, in order to guarantee that the thermal demand can be fulfilled at all times at an admissible temperature level. At the same time, the following *optimization targets* should be considered:

- The local consumption of on-site PV production for thermal production should be maximized.
- On-site CO_2 emissions and electricity imported from the external grid should be minimized.

These operational targets can be translated into an operational strategy, which was implemented as model-based optimal control (compare with Section 4.4.2), based on a linear optimization problem formulation. It prioritizes the heat sources according to the following scheme:

1. In case there is sufficient PV production, heat pumps are given priority over all other heat producers in order to maximize the consumption of on-site PV production.
2. In case the heat pumps cannot provide sufficient generation, the BB is used.
3. In case heat pumps and the BB combined cannot provide sufficient generation, the CHP is used.
4. In case the demand is still not met, the GBs are fired.

At the same time, the controller keeps track of all operational constraints (matching of generation and demand, production thresholds, network constraints, etc.). This guarantees the optimal operation for a given system configuration.

6.1.3. Design Optimization Strategy

For optimizing the system design, the heating capacities of the two heat pumps are the *degrees of freedom*. However, only a limited number of realistic sizing options has been identified beforehand by the owner, based on experience and expertise from domain experts as well as spatial and budget constraints. For instance, since WWHPs are more efficient than GWHPs, operation of the WWHP is prioritized, whereas the GWHP is only turned on if even more electricity from PV is available and heat is needed. Therefore, all selected configurations foresee a bigger WWHP in combination with a smaller GWHP. Furthermore, given the operational strategy explained above, which also aims at a minimization of the electricity consumption from the external grid, the maximal practical size of the heat pumps is limited by the maximal PV production. These considerations led to three potential system configurations (see Table 3), which differ in the sizing of the heat pumps.

Table 3. Considered system configurations.

Configuration Name	WWHP Size (kW_{el})	GWHP Size (kW_{el})
A	100	50
B	150	50
C	200	50

With the limited number of system configurations and the possibility to translate the operational strategy into an optimal controller scheme, the OCS is the natural choice of *optimization approach* for this application (see Figure 6).

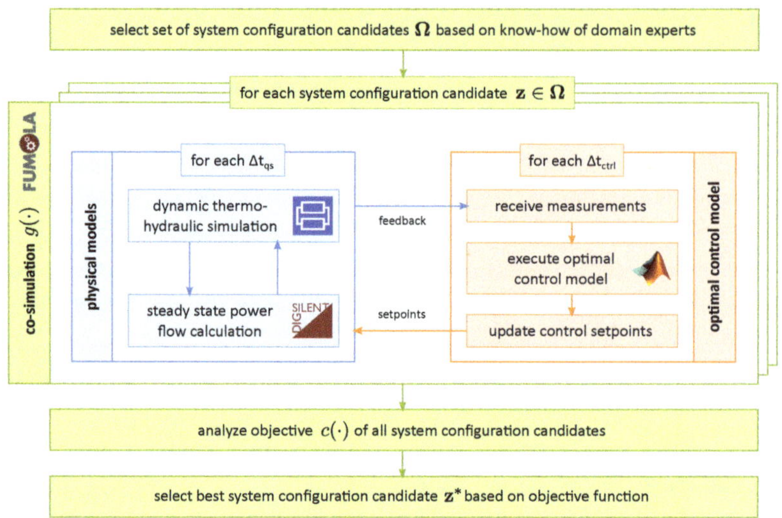

Figure 6. Schematic overview of the OCS workflow applied to the system design of the suburban industrial area.

6.1.4. Results

As expected, the operational strategy is affected by the seasonal variations (temperature, solar irradiance), effectively resulting in seasonal operational modes that prefer different sources at different times of the year. These significant differences necessitate the evaluation of the system performance on a yearly basis, in order to provide a good basis for the choice of the optimal design.

Figure 7 summarizes the performance of the three considered system configurations in terms of yearly total energy production for the heat pumps (E_{tot}^{HP}), the biomass boiler (E_{tot}^{bio}) and the CHP (E_{tot}^{CHP}). In all three cases, the thermal energy production of the gas boilers is around 230 MWh_{th}/a. The figure shows that Configuration C—which has the largest WWHP—fails to exploit the full potential of the PV generation, resulting in the smallest total energy production of all configurations. This is due to the fact that the heat pump must not be operated below 80% of its maximum capacity, which in turn leads to a high threshold for turning it on in Configuration C. Configuration A—with the smallest WWHP—slightly falls behind Configuration B, which provides in this regard the best compromise as it exhibits the largest total generation from the heat pumps. Furthermore, in Configuration C, the WWHP's high operational threshold causes not only an increase of biomass-based generation, but actually leads to an overcompensation at the cost of the CHP-based thermal generation compared to the other configurations.

Analysis of the detailed results from the co-simulation shows that Configuration B has the most favorable impact on the electrical system in view of integrating the on-site PV production. For instance, it shows the most significant improvement of the voltage band usage compared to non-hybrid system layouts by reducing the time and amount the maximum voltage band exceeds the 10% threshold.

In conclusion, even though the improvements from the system level point of view are limited due to the actual available PV generation and other practical constraints (e.g., economical aspects of increasing the CHP capacity), Configuration B shows overall the best thermal and electrical system performance. It optimally exploits the on-site PV generation via the heat pumps ($E_{tot}^{HP} = 225\,MWh_{th}/a$) while at the same time providing a modest reduction of fuel-based generation from the CHP and the biomass boiler ($\Delta E_{tot}^{fuel} = -211\,MWh_{th}/a$). Furthermore, the design goal of minimizing on-site CO_2 is successfully met through the preference for the heat pumps, the biomass boiler and the CHP over the gas boiler (see Figure 8).

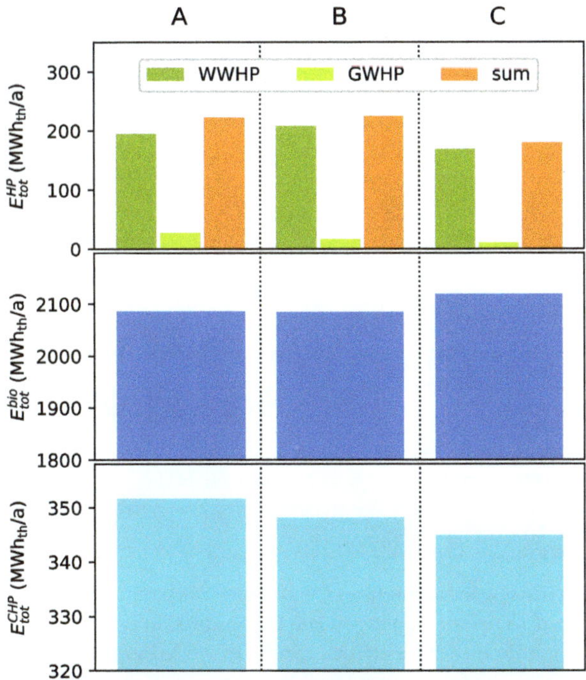

Figure 7. Comparison of the yearly thermal energy production of heat pumps, biomass boiler and CHP for the three considered system configurations (Configuration A, Configuration B and Configuration C).

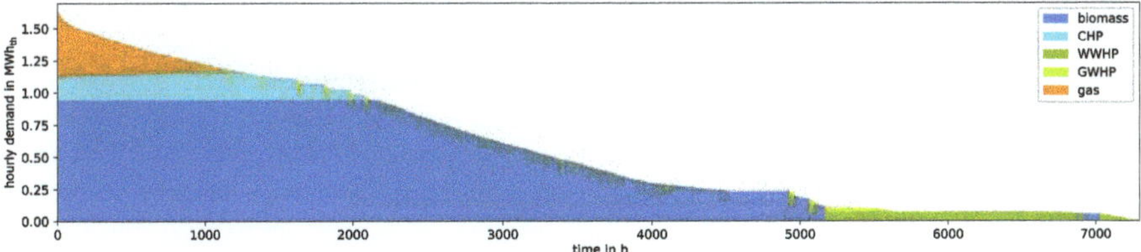

Figure 8. Yearly load duration curve.

6.2. HPS Example Application: Rural Residential Area

6.2.1. Technical System Layout

The system comprises a rural low-voltage electric grid with a total cable length of about 6.5 km. This network connects residential, commercial and agricultural customers, summing up to a total number of around 110 customers with a total annual power demand of around 775 MWh_{el} and PV systems that are feeding a total of 384 MWh_{el} per year.

The district heating network structure is typical for rural areas with low heat demand density (linear density of about 520 $kWh_{th}/a/m$ and peak load demand of about 2.0 MW_{th}). An outdoor-temperature dependent heating curve is used to set the supply temperature between 90 (winter) and 70 °C (summer). Return and supply pipes connect around 60% of the buildings in the area with a base heat generation plant responsible to keep the differential pressures at the consumer substations above a minimum.

Additionally, three electric boilers, i.e., electric heaters combined with thermal storage tanks, are installed as *coupling points* between the networks. Adequate locations for the boilers were identified using a simulation-based sensitivity analysis in an early planning phase. These locations are especially prone to over voltage problems resulting from PV generation, an issue that might be eased by the active conversion of excess power generation. The volume of the thermal storage tanks and the capacities of the electric heaters are chosen as *degrees of freedom* of this system. The system layout is illustrated in Figure 9.

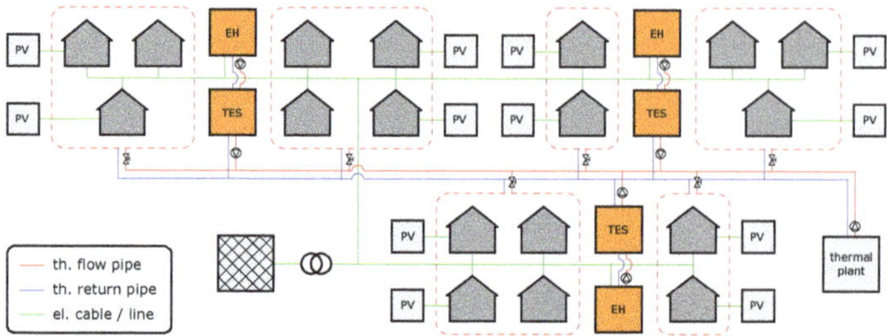

Figure 9. Schematic overview of the hybrid system layout for the rural residential area, depicting the electrical distribution network and the district heating network (EH, electric heater; TES, thermal energy storage). The components targeted by the design optimization process are highlighted in orange.

6.2.2. Operational Strategy

Decentralized electricity generation from PV systems of local prosumers pose challenges to the electrical distribution grid. Especially in times when PV production is high and electrical consumption is low, the upper voltage limit in the network can be exceeded and lines or transformers can be overloaded. To mitigate some of these problems, the operational strategy foresees to utilize excess power from PV overproduction. The local power grid is regarded as virtual power plant (VPP) and a supervisory controller tries to use as much of the excess power, i.e., negative residual load, locally via the installed electric heaters. Based on these ideas, a suitable operational strategy was implemented in Python using the following rule-based scheme (compare with Section 4.4.1):

1. In case the VPP generates excess power, the electric heaters are set to utilize this power and store it in the respective storage tanks as long as their temperatures are below 95 °C.
2. The thermal storage tanks are discharged as long as the temperatures in their top layers are above the current district heating supply temperature and only if there is enough district heating demand.

6.2.3. Design Optimization Strategy

The high number of possible system configurations and the use of a rule-based operational strategy makes HPS the most appropriate *optimization approach* for this application. Figure 10 illustrates the overall simulation-based optimization procedure.

The design optimization of the coupling units, i.e., finding optimal sizes for the defined *degrees of freedom*, rests upon the following *optimization targets*:

- Reducing heat generation from the main supply unit in the district heating network is rewarded.
- Costs for storage tanks and electric heater capacities linearly increase depending on size.
- Potentially increased power imports into the electric network introduced by the electric heaters are penalized.

These targets are combined into a single objective function using appropriate weighting factors. The six-dimensional solution space is reduced by introducing bound constraints for the degrees of freedom, avoiding extremely high and low tank volumes and electric heater capacities.

The implementation relies on the use of a dedicated open-source tool for simulation-based optimization [46], which allows parallelization on multiple machines and the use of optimization algorithms specialized for efficient black-box optimization. In this case, the freely available PSwarm solver [45] is used that combines particle swarm optimization and pattern search for efficient global optimization. The pattern search relies on a coordinate search method that is responsible for local convergence, whereas the population-based particle swarm algorithm performs a global search enabling the exploration of the whole design space. The stopping criterion is based on a maximum number of co-simulation runs, i.e., objective function calls.

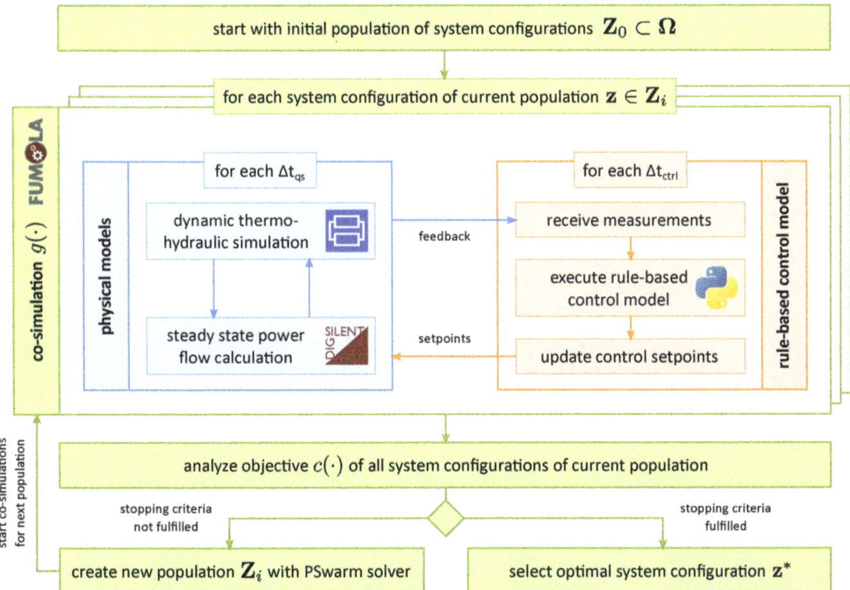

Figure 10. Schematic overview of the HPS workflow applied to the system design of the rural residential area.

6.2.4. Results

The convergence of the objective function value over the number of co-simulation runs, i.e., assessed system configurations, is shown in Figure 11. The objective function values are shown relative to a reference solution reflecting the current status quo, i.e., without any active coupling between the networks using the electric heaters. The optimization algorithm is able to outperform the reference scenario within only a few simulation runs. After around 450 runs, the algorithm converges to a minimum and stops after it exceeds the maximum number of objective function calls. Although this might only be a local minimum, the found minimum objective function is around 5.7% lower compared to the uncoupled case. The corresponding optimal system configuration exhibits electric heaters with a total capacity of 260 kW_{el} and thermal storage tanks with a total volume of 20 m^2.

The impact on the two networks for the uncoupled case and the optimal design case in terms of load duration curves is shown in Figure 12. A total of around 220 MWh electricity is converted in the optimal design case using electric boilers, illustrated by the colored areas. It can be seen that electric heaters in combination with thermal storage tanks are able to use a significant share of the excess power generation from PV by converting it into heat that is fed to the district heating network. Due to the seasonality of PV generation, district heating is mainly affected in low heat demand times, i.e., summer.

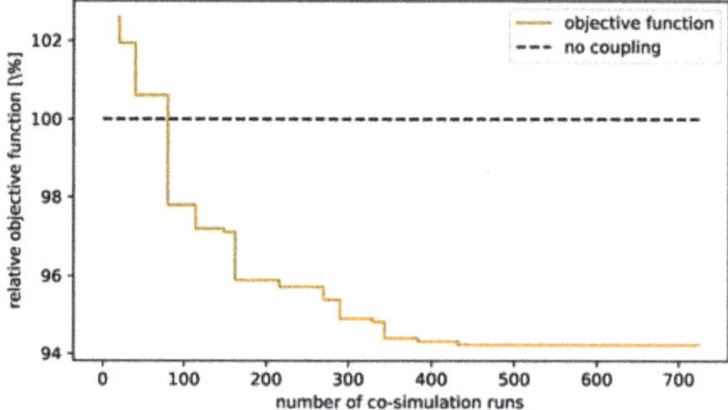

Figure 11. Convergence of objective function shown relative to the reference scenario without coupling between the networks, i.e., no electric boilers.

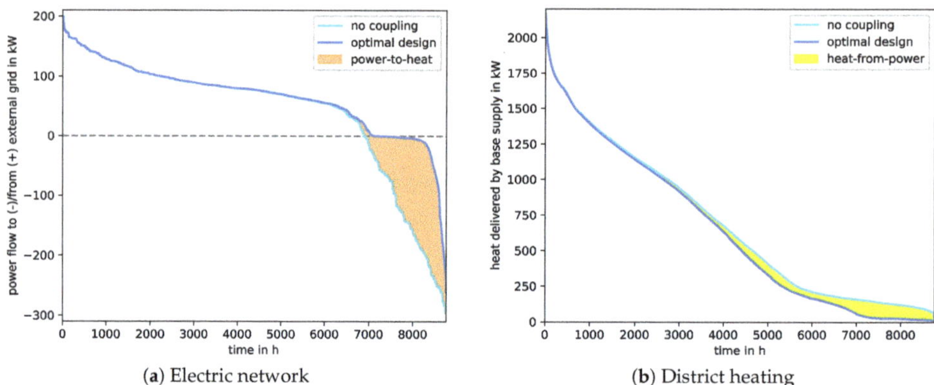

Figure 12. Load duration curves for the electric network (**a**) and the district heating network (**b**) for the uncoupled and the optimal design system configurations.

7. Conclusions

This work presents a simulation-based design approach for hybrid thermal-electrical distribution grids. The approach addresses the technical challenges of both domains while at the same time emphasizing their mutual control and operation, in order to exploit hitherto unused synergies in production, storage and consumption. As such, this approach is a significant extension of established tools for the design optimization of multi-energy systems.

Co-simulation is recommended for the technical assessment as a means to bridge the gap between single-domain simulation tools and the multi-domain target of investigation, providing a viable and practical approach to involve experts from different domains. A novel methodology is presented that links complementary design approaches to suitable operational strategies and optimization methods according to the state of the art. This enables the exploitation of synergies in the control and operation of hybrid thermal-electrical distribution systems in an optimal way. Furthermore, an implementation of the proposed approach based on state-of-the-art tools is presented. Finally, the applicability of the presented approach is demonstrated in two real-world applications.

Future work could apply this approach to other network-related multi-energy applications, such as design of short-term and long-term storage, peak shaving and others. A method for representative days selection for coupled heat and power networks could avoid the need for full-year simulations and, thus, reduce computational time. The use of computationally efficient model-based optimal control models or faster models for the physical system could enable the combination with meta-heuristic design optimization in a tolerable run time.

Author Contributions: conceptualization, E.W. and B.L.; methodology, E.W. and B.L.; software, E.W., B.L., D.B., S.H. and T.F.; investigation, E.W., B.L., D.B., S.H. and T.F.; writing—original draft preparation, E.W. and B.L.; writing—review and editing, D.B., S.H. and R.H.; visualization, E.W. and B.L.; supervision, R.H. and E.W.; project administration, E.W. and B.L.; and funding acquisition, E.W. All authors have read and agreed to the published version of the manuscript.

Funding: Parts of this work were funded by the Austrian research funding association (FFG) within the scope of the research project *OptHySys—Optimierung Hybrider Energienetze und -Systeme* (project #848778). Parts of this work were carried out within the research project *SmILES—Smart Integration of Energy Storages in Local Multi Energy Systems for maximising the Share of Renewables in Europe's Energy Mix* and received funding from the European Commission's Horizon 2020 Research and Innovation Programme under the Grant Agreement No. 730936.

Acknowledgments: The authors want to acknowledge support provided by the cooperation college *Smart Industrial Concept* (SIC!).

Conflicts of Interest: The authors declare no conflict of interest. The funders had no role in the design of the study; in the collection, analyses, or interpretation of data; in the writing of the manuscript, or in the decision to publish the results.

Abbreviations

The following abbreviations are used in this manuscript:

API	application programming interface
BB	biomass boiler
CHP	combined heat and power
DH	district heating
FMI	Functional Mock-up Interface
FMU	Functional Mock-up Unit
GB	gas boiler
GWHP	ground water heat pump
MPC	model predictive control
PV	photovoltaic
RES	renewable energy sources
VPP	virtual power plant
WWHP	waste water heat pump

References

1. Mancarella, P.; Andersson, G.; Pecas-Lopes, J.A.; Bell, K.R.W. Modelling of integrated multi-energy systems: Drivers, requirements, and opportunities. In Proceedings of the Power Systems Computation Conference (PSCC), Genoa, Italy, 20–24 June 2016; pp. 1–22.
2. Van Beuzekom, I.; Gibescu, M.; Slootweg, J.G. A review of multi-energy system planning and optimization tools for sustainable urban development. In Proceedings of the IEEE Eindhoven PowerTech, Eindhoven, The Netherlands, 29 June–2 July 2015; pp. 1–7.
3. Sustainable Energy Planning Research Group at Aalborg University in cooperation with PlanEnergi and EMD A/S. EnergyPLAN. Available online: https://www.energyplan.eu (accessed on 29 February 2020).
4. HOMER Energy LLC. HOMER. Available online: https://www.homerenergy.com (accessed on 29 February 2020).
5. Palensky, P.; Widl, E.; Elsheik, A. Simulating Cyber-Physical Energy Systems: Challenges, Tools and Methods. *Syst. Man Cybern. Syst. IEEE Trans.* **2014**, *44*, 318–326. [CrossRef]
6. Fritzson, P. *Introduction to Modeling and Simulation of Technical and Physical Systems with Modelica*, 1st ed.; Wiley-IEEE Press: Hoboken, NJ, USA, 2011.
7. Franke, R.; Wiesmann, H. Flexible modeling of electrical power systems—The Modelica PowerSystems library. In Proceedings of the of the 10th International Modelica Conference, Lund, Sweden, 10–12 March 2014; pp. 515–522.
8. Vanfretti, L.; Rabuzin, T.; Baudette, M.; Murad, M. iTesla Power Systems Library (iPSL): A Modelica library for phasor time-domain simulations. *SoftwareX* **2016**, *5*, 84–88. [CrossRef]

9. Wetter, M.; Zuo, W.; Nouidiu, T.S.; Pang, X. Modelica Buildings library. *J. Build. Perform. Simul.* **2014**, *7*, 253–270. [CrossRef]
10. Bonvini, M.; Wetter, M. Gradient-based optimal control of batteries and HVAC in district energy systems. In Proceedings of the 14th Conference of International Building Performance Simulation Association (BS2015), Hyderabad, India, 7–9 December 2015; pp. 363–370.
11. The Mathworks. MATLAB and Simulink. Available online: https://www.mathworks.com (accessed on 29 February 2020).
12. Leitner, B.; Widl, E.; Gawlik, W.; Hofmann, R. A method for technical assessment of power-to-heat use cases to couple local district heating and electrical distribution grids. *Energy* **2019**, *182*, 729–738. [CrossRef]
13. Wetter, M. Co-simulation of building energy and control systems with the Building Controls Virtual Test Bed. *J. Build. Perform. Simul.* **2011**, *4*, 185–203. [CrossRef]
14. Widl, E.; Delinchant, B.; Kübler, S.; Li, D.; Müller, W.; Norrefeldt, V.; Nouidui, T.S.; Stratbücker, S.; Wetter, M.; Wurtz, F.; et al. Novel simulation concepts for buildings and community energy systems based on the Functional Mock-up Interface specification. In Proceedings of the Workshop on Modeling and Simulation of Cyber-Physical Energy Systems (MSCPES), Berlin, Germany, 14 April 2014; pp. 1–6.
15. Georg, H.; Wietfeld, C.; Müller, S.C.; Rehtanz, C. A HLA Based Simulator Architecture for Co-simulating ICT Based Power System Control and Protection Systems. In Proceedings of the 3rd IEEE International Conference on Smart Grid Communications (SmartGridComm), Tainan, Taiwan, 5–8 November 2012.
16. Rohjans, S.; Lenhoff, S.; Schütte, S.; Scherfke, S.; Hussain, S. Mosaik—A modular platform for the evaluation of agent-based Smart Grid control. In Proceedings of the IEEE PES ISGT Europe, Lyngby, Denmark, 6–9 October 2013; pp. 1–5.
17. Galtier, V.; Vialle, S.; Dad, S.; Tavella, J.-P. Lam-Yee-Mui, J.-P.; Plessis, G. FMI-Based Distributed Multi-Simulation with DACCOSIM. In Proceedings of the Symposium on Theory of Modeling and Simulation (TMS'15), Alexandria, VI, USA, 12–15 April 2015; pp. 804–811.
18. Widl, E.; Müller, W.; Basciotti, D.; Henein, S.; Hauer, S.; Eder, K. Simulation of multi-domain energy systems based on the Functional Mock-up Interface specification. In Proceedings of the International Symposium on Smart Electric Distribution Systems and Technologies (EDST), Vienna, Austria, 8–11 September 2015; pp. 510–515.
19. Widl, E.; Jacobs, T.; Schwabender, D.; Nicolas, S.; Basciotti, D.; Henein, S.; Noh, T.-G.; Terreros, O.; Schuelke, A.; Auer, H. Studying the potential of multi-carrier energy distribution grids: A holistic approach. *Energy* **2018**, *153*, 519–529. [CrossRef]
20. Geidl, M.; Andersson, G. Optimal Power Flow of Multiple Energy Carriers. *IEEE Trans. Power Syst.* **2007**, *22*, 145–155. [CrossRef]
21. Walker, S.; Labeodan, T.; Maassen, W.; Zeiler, W. A review study of the current research on energy hub for energy positive neighborhoods. In Proceedings of the CISBAT 2017 International Conference, Lausanne, Switzerland, 6–8 September 2017, doi:10.1016/j.egypro.2017.07.387. [CrossRef]
22. Wetter, M.; Wright, J. A comparison of deterministic and probabilistic optimization algorithms for nonsmooth simulation-based optimization. *Build. Environ.* **2004**, *39*, 989–999, doi:10.1016/j.buildenv.2004.01.022. [CrossRef]
23. Huang, Y.; lei Niu, J. Optimal building envelope design based on simulated performance: History, current status and new potentials. *Energy Build.* **2016**, *117*, 387–398. [CrossRef]
24. Niemelä, T.; Kosonen, R.; Jokisalo, J. Cost-optimal energy performance renovation measures of educational buildings in cold climate. *Appl. Energy* **2016**, *183*, 1005–1020. [CrossRef]
25. Delgarm, N.; Sajadi, B.; Kowsary, F.; Delgarm, S. Multi-objective optimization of the building energy performance: A simulation-based approach by means of particle swarm optimization (PSO). *Appl. Energy* **2016**, *170*, 293–303. [CrossRef]
26. Nguyen, A.T.; Reiter, S.; Rigo, P. A review on simulation-based optimization methods applied to building performance analysis. *Appl. Energy* **2014**, *113*, 1043–1058. [CrossRef]
27. Wang, H.; Wang, H.; Zhou, H.; Zhu, T. Modeling and optimization for hydraulic performance design in multi-source district heating with fluctuating renewables. *Energy Convers. Manag.* **2018**, *156*, 113–129. [CrossRef]

28. van der Heijde, B.; Vandermeulen, A.; Salenbien, R.; Helsen, L. Integrated Optimal Design and Control of Fourth Generation District Heating Networks with Thermal Energy Storage. *Energies* **2019**, *12*, 2766. [CrossRef]
29. Widl, E. FUMOLA–Functional Mock-up Laboratory. Sourceforge.net. Available online: http://sourceforge.fumola.net (accessed on 29 February 2020).
30. Blochwitz, T.; Otter, M.; Arnold, M.; Bausch, C.; Elmqvist, H.; Junghanns, A.; Maus, J.; Monteiro, M.; Neidhold, T.; Neumerkel, D.; et al. The Functional Mockup Interface for Tool independent Exchange of Simulation Models. In Proceedings of the 8th International Modelica Conference, Dresden, Germany, 20–22 March 2011.
31. Dassault Systèmes. DYMOLA Systems Engineering. Available online: https://www.3ds.com/products-services/catia/products/dymola (accessed on 29 February 2020).
32. Leitner, B. Modelica DisHeatLib library. GitHub.com. Available online: https://github.com/AIT-IES/DisHeatLib (accessed on 29 February 2019).
33. Modelica IBPSA Library. Available online: https://github.com/ibpsa/modelica-ibpsa (accessed on 21 March 2020).
34. Kusuda, T.; Achenbach, R.P. Earth temperature and thermal diffusivity at selected stations in the United States. *ASHRAE Trans.* **1965**, *71*, 61–75.
35. Colebrook, C.F. Turbulent flow in pipes, with particular reference to the transition region between the smooth and rough pipe laws. *J. Inst. Civ. Eng.* **1939**, *11*, 133–156. [CrossRef]
36. Van der Heijde, B.; Fuchs, M.; Ribas Tugores, C.; Schweiger, G.; Sartor, K.; Basciotti, D.; Müller, D.; Nytsch-Geusen, C.; Wetter, M.; Helsen, L. Dynamic equation-based thermo-hydraulic pipe model for district heating and cooling systems. *Energy Convers. Manag.* **2017**, *151*, 158–169. [CrossRef]
37. Bergman, T.; Lavine, A.; Incropera, F.; DeWitt, D. *Fundamentals of Heat and Mass Transfer*, 8 ed.; John Wiley & Sons Inc.: Hoboken, NJ, USA, 2018.
38. Kleinbach, E.; Beckman, W.; Klein, S. Performance study of one-dimensional models for stratified thermal storage tanks. *Sol. Energy* **1993**, *50*, 155–166. [CrossRef]
39. DIgSILENT PowerFactory. DIgSILENT PowerFactory. Available online: http://www.digsilent.com (accessed on 29 February 2020).
40. Widl, E.; Leitner, B. The FMI++ PowerFactory FMU Export Utility. Sourceforge.net. Available online: http://powerfactory-fmu.sourceforge.net (accessed on 29 February 2020).
41. Widl, E. The FMI++ MATLAB Toolbox for Windows. Sourceforge.net. Available online: http://matlab-fmu.sourceforge.net (accessed on 29 February 2020).
42. Widl, E. FMI++ Python Interface. Python Package Index. Available online: https://pypi.org/project/fmipp (accessed on 29 February 2020).
43. Eberhart, R.C.; Kennedy, J. A new optimizer using particle swarm theory. In Proceedings of the Sixth International Symposium on Micromachine and Human Science, Nagoya, Japan, 4–6 October 1995; pp. 39–43.
44. Storn, R.; Price, K. Differential Evolution: A Simple and Efficient Adaptive Scheme for Global Optimization over Continuous Spaces. *Glob. Optim.* **1997**, *11*, 341–359. [CrossRef]
45. Vaz, A.; Vicente, L. PSwarm: A hybrid solver for linearly constrained global derivative-free optimization. *Optim. Methods Softw.* **2009**, *24*, 669–685. [CrossRef]
46. Pesendorfer, B.; Widl, E. The optFUMOLA package: A simulation-based black-box optimization library and interface. In Proceedings of the Workshop on Modeling and Simulation of Cyber-Physical Energy Systems (MSCPES), Pittsburgh, PA, USA, 21 April 2017; pp. 1–6.

Article

Vehicle-To-Grid for Peak Shaving to Unlock the Integration of Distributed Heat Pumps in a Swedish Neighborhood

Monica Arnaudo *, **Monika Topel and Björn Laumert**

Royal Institute of Technology, 11428 Stockholm, Sweden; monika.topel@energy.kth.se (M.T.); bjorn.laumert@energy.kth.se (B.L.)

* Correspondence: monica.arnaudo@energy.kth.se

Received: 17 February 2020; Accepted: 2 April 2020; Published: 3 April 2020

Abstract: The city of Stockholm is close to hitting the capacity limits of its power grid. As an additional challenge, electricity has been identified as a key resource to help the city to meet its environmental targets. This has pushed citizens to prefer power-based technologies, like heat pumps and electric vehicles, thus endangering the stability of the grid. The focus of this paper is on the district of Hammarby Sjöstad. Here, plans are set to switch from district heating to heat pumps. A previous study verified that this choice will cause overloadings on the electricity distribution grid. The present paper tackles this problem by proposing a new energy storage option. By considering the increasing share of electric vehicles, the potential of using the electricity stored in their batteries to support the grid is explored through technical performance simulations. The objective was to enable a bi-directional flow and use the electric vehicles' (EVs)' discharging to shave the peak demand caused by the heat pumps. It was found that this solution can eliminate overloadings up to 50%, with a 100% EV penetration. To overcome the mismatch between the availability of EVs and the overloadings' occurrence, the minimum state of charge for discharging should be lower than 70%.

Keywords: vehicle-to-grid; heat pumps; integrated energy systems

1. Introduction

Sweden has adopted the ambitious goal of becoming a zero carbon emission society by 2045 [1]. Within this context, the energy sector has attracted a special attention because it accounts for more than 70% (including transport) of total greenhouse gas emissions [2]. When looking at the internal electricity mix of this country [3,4], electricity is promoted as a clean resource and thus a promising option to achieve the 2045 target. As a consequence, in Stockholm, citizen-driven initiatives work to promote the installation of distributed residential heat pumps (HPs) and the adoption of electric vehicles (EVs) [5]. A real example is represented by Hammarby Sjöstad, which is a residential neighborhood constituted by multi-apartment buildings. This district is currently connected to the city's district heating (DH) network, but projects are already approved for some customers to switch to their own domestic HP.

On an opposite side, this electrification trend is challenged by a structural problem that impacts the power grid system in Stockholm. In fact, the electricity transmission to the city is close to its maximum power capacity. This means that the distribution system operators (DSOs) are being forced to impose power limits on the distribution grid [6]. Thus, the connection of new loads, like HPs and EVs, could soon be impossible. If capacity expansion investments are to be avoided or delayed, alternative solutions should be quickly found.

A promising option is offered by energy storage [7]. This technology can shift loads in time by charging when surplus energy supply is available and discharging during periods of peak demand.

This can be done on both the heat and on the electricity sides, depending on the type of energy storage technology used.

For this paper, the focus was set on the electric power side by considering the electricity storage potential in the batteries of EVs. In this context, an EV is regarded not only as a load to the power grid but also as an extended capacity for supply. This is based on the vehicle-to-grid (V2G) concept. V2G technology enables a bi-directional flow between an EV's battery and the grid [8–11]. Thus, both the charging and discharging of electrical energy are allowed.

The background of this study was linked to a real initiative called "Charge at Home" ("Ladda Hemma" in the original Swedish). Within this context, several housing associations in Hammarby Sjöstad were encouraged and assisted in the process of installing EV charging infrastructures in their buildings. The study presented in this paper explores the potential of enabling a V2G bi-directional flow at these stations. By assuming that distributed domestic HPs are installed in Hammarby Sjöstad, the objective was to use the electricity stored in the EVs' batteries to cover the peak power demand generated by these HPs. The main challenges were the availability of EVs parked at the stations and the level of charge of their batteries.

The importance of this study was stressed by a previous assessment that the installation of distributed HPs in Hammarby Sjöstad will cause overloadings on the local electricity distribution grid [12]. This was found by assuming that DSOs would impose a maximum loading limit equal to 100% for each grid's cable. The possibility of using the thermal mass of the buildings as thermal energy storage (TES) by controlling the thermostats of the apartments was also studied. Despite the implementation of this solution, grid overloadings were still detected. The present work builds on this last outcome by verifying how much V2G technology can further contribute to balance grid overloadings.

Several papers have analyzed the possibility of using V2G for peak shaving and avoiding grid upgrades. Most of them [13–18] proposed optimal control algorithms by looking at costs and emissions. Some of them also took battery degradation [19,20], market composition [21], and incentives [22] into account. The studies of [23,24] combined the capacity provision market with the spinning reserve. The authors of [25] showed the advantage of combining a V2G peak shaving strategy with a domestic battery and a photovoltaic (PV) system.

Despite the broad perspective, none of these papers considered the potential synergies and impacts between the power, transport, and heating sectors.

The authors of [26] implemented V2G with a time-of-use tariff. This study also showed the impact of considering the indoor temperature of buildings as a constraint on the supply dispatch of a micro-grid. However, their perspective was limited to one building.

The energy hub presented in [27] included the heating and cooling demands of a residential community. How the implementation of V2G can impact electricity and cooling prices was shown. However, impact and synergies among these energy carries were not considered from a technical standpoint.

Given these literature gaps, one objective of this paper was to propose a district-level perspective on potential synergies among the heating, electricity, and transport sectors. To approach this sector-coupling problem, the co-simulations of dedicated models [12,28] and stochastic profiles were combined. Thus, the technical performances of technologies belonging to these different sectors were linked.

Here, the focus was on the case of a multi-apartments neighborhood located in Stockholm, Hammarby Sjöstad. By taking into account that distributed HPs can overload the local grid (reference scenario from [12] as a main objective, this paper aimed at estimating the potential of V2G to alleviate this problem (V2G integration scenario). This was done by using the energy stored in the EV's batteries to shave the HPs' peak power demand.

The remainder of the paper is as follows. Firstly, the case study is presented in Section 2. Secondly, the two scenarios and the corresponding simulation models are described in Sections 3 and 4, respectively. Finally, the results are discussed in Section 5.

2. Case Study

The case study presented in this paper corresponded to a specific area in Hammarby Sjöstad, as shown in Figure 1. All the buildings in this area are connected to a single medium-to-low voltage transformer substation. The conceptual illustration of the case study area in Figure 1 sketches the current type of energy infrastructure in the neighborhood. The multi-apartment buildings are connected to a DH network for space heating and domestic hot water purposes. The installed capacity is about 1.2 MW for a total of 94,555 m^2 of heated area. Electricity is provided by a low voltage distribution grid (400 V) that covers a capacity of about 1.6 MW.

Figure 1. The case study area: a conceptual illustration and a satellite view BB: building block; DH: district heating.

Within the scope of the scenarios later described in Section 3, the buildings are grouped in 22 building blocks (BB) according to the energy declaration documents of the corresponding housing associations [29]. As reported in Table 1, the BBs are identified through their heated area.

Table 1. Square meters of heated area for each BB.

BB1	BB2	BB3	BB4	BB5	BB6	BB7	BB8	BB9	BB10	BB11
4576	4054	919	3258	4889	8405	1613	1613	2831	1031	3138
BB12	**BB13**	**BB14**	**BB15**	**BB16**	**BB17**	**BB18**	**BB19**	**BB20**	**BB21**	**BB22**
1718	1931	3340	5021	15,060	3466	10,699	3632	3560	3560	6241

The choice of a limited area was motivated by the focus of the study, which was to show the potential of peak power demand management by using V2G technology. This could be easily replicated to show the impact on the whole neighborhood (or beyond).

3. Scenarios

3.1. Reference Scenario

The reference scenario for the present work was taken from a previous study presented in [12]. Given the interest of a few housing associations (Section 1), the focus was to assess the technical feasibility of replacing a DH network with domestic-distributed HPs. Since the sole installation of the distributed HPs was shown to cause overloadings on the electricity distribution grid, a demand response solution was further assessed. The thermal mass of the buildings was used as TES to alleviate the grid from the overloadings caused by the HPs. A positive contribution was shown with a reduction of the heat demand dependence from a DH of 6%, compared to the case without the thermal mass control. Furthermore, up to 50% fewer overloadings were detected at the HP level. However, it was also highlighted that a full disconnection from the DH network was not possible given the current infrastructure capacity. These outcomes [12] represented the reference scenario for the present study.

3.2. V2G Integration Scenario

In the present paper, the reference scenario was extended by assessing the potential of V2G technology as a way to further support the electricity distribution grid. A simplified illustration of this concept is shown in Figure 2. For the sake of simplicity, a BB is represented as one building, though it can also correspond to a group of buildings. The V2G charging/discharging station is located upstream of each HP installation. The EVs' charging/discharging patterns were based on stochastic profiles. These were generated according to the assumptions described in Section 4.

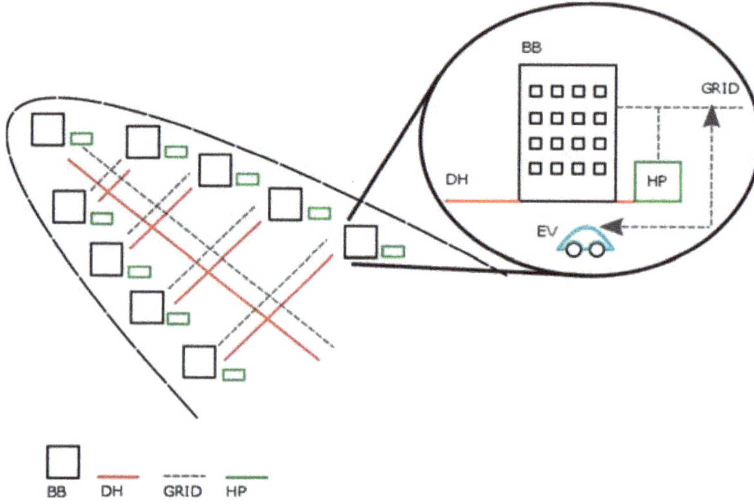

Figure 2. Conceptual illustration of the case study within the vehicle-to-grid (V2G) scenario. HP: heat pump; EV: electric vehicle.

The EVs' parking stations are located in the BBs, in line with the "Charge at Home" initiative (Section 1). In this scenario, as illustrated in Figure 2, both charging and discharging processes—and thus V2G—were enabled in each BB. The aim was to cover the grid overloadings by performing peak shaving with the electricity stored in the EVs' batteries. This study considered a 100% penetration of the EVs. Furthermore, the EVs' charging was assumed to happen overnight with no overloadings. More specific assumptions are described in Section 4.

4. Models

Different combinations of models were used to represent the two scenarios described in Section 2. These models are discussed in detail in the following sub-sections.

4.1. Models for the Reference Scenario

This sub-section briefly highlights that, from a modeling and simulation perspective, a co-simulation method [28,30] was used to assess the interaction between the loading of the electricity distribution grid, the heat supply from HPs and a DH network, and the indoor temperature of residential buildings. Figure 3 sums up the modeling tools that were used for each system involved in the scenarios. The scheme also shows the main heat (red lines), power flows (black lines), and control signals (blue dashed lines).

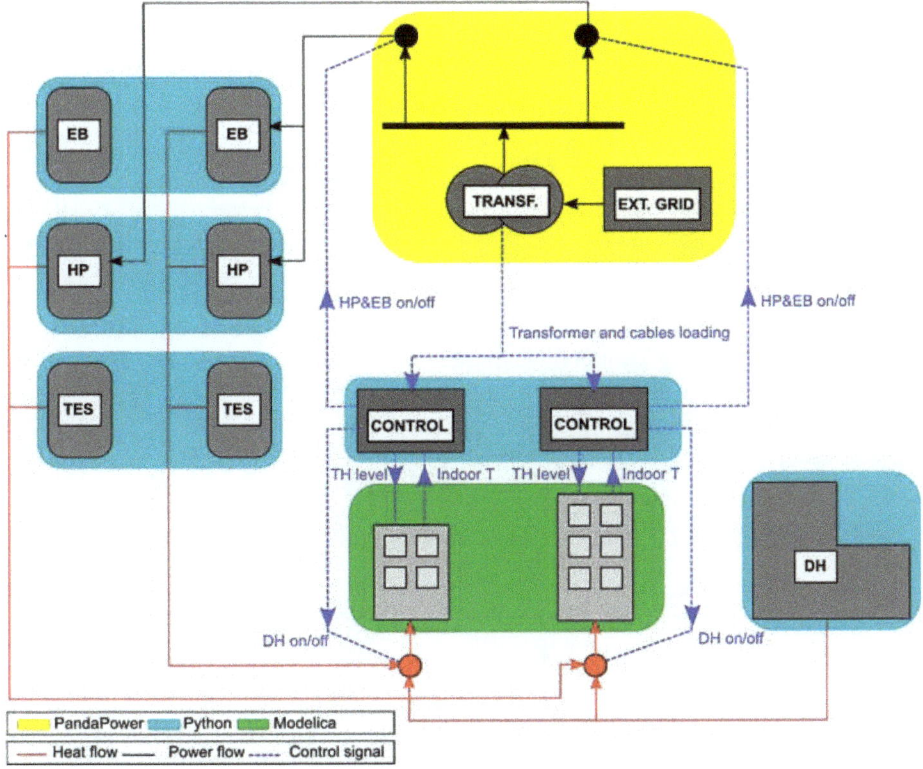

Figure 3. Co-simulation scheme for the reference scenario (EB: electrical backup; TH: thermostat).

The co-simulation environment allowed us to capture two feedback loop signals. On one hand, the loading status of the grid was checked to decide whether the HPs could be operated. An overloading signal was generated when 100% of the capacity was reached. On the other hand, the indoor temperature of the BBs was monitored to decide if the thermostat set point could be changed to load/unload the thermal mass. In this case, the indoor temperature was to be maintained within the range of 20 ± 0.5 °C.

The following models were implemented for this reference scenario:

- A power flow model for the electricity distribution grid. An open source tool called Pandapower [31] was used for this purpose.
- A reduced order model for the thermal demand of the BBs. This model is part of the Modelica Buildings library [32], maintained by the International Building Performance Association (IBPSA).
- A Python model based on a coefficient of performance formula [33] for the HPs.

- Python models based on energy balance equations for the DH, the TES, and the electrical backup units.
- A Python in-house operation logic for the control of the heating supply.

These models were used within a one year simulation with an hourly time step. A detailed explanation, parameters, variables, and assumptions for all the models are provided in [12].

4.2. Models for the V2G Integration Scenario

Concerning the V2G integration scenarios, two models were combined:

- The grid overloading power profiles generated as a result of the reference scenario for the critical cables.
- Stochastic driving patterns for the EVs owned by the inhabitants of Hammarby Sjöstad, assuming a 100% penetration of this technology.

Regarding the first model, a critical day was selected (10th January) from the reference scenario, together with the identified critical cables. With reference to this day, the overloading power for each critical cable was obtained by subtracting the power corresponding to a 100% loading of each critical cable from the actual loading. For the hours that did not present overloadings, the overloading power was set to zero.

In this study, the analysis was extended to all cables, including the parallel ones. Figure 4 shows the one-line diagram of the electricity distribution grid of the case study. The four red-highlighted cables are the ones that were identified as critical in the reference scenario. Two of these cables corresponded to bundled parallel cables. The corresponding labels are listed as the following:

- Cable 1—LU 1
- Cable 2—Llugnw24, Llugnw26, Llugnw27
- Cable 3—Luddnw3
- Cable 4—Llugnw6, Llugnw7

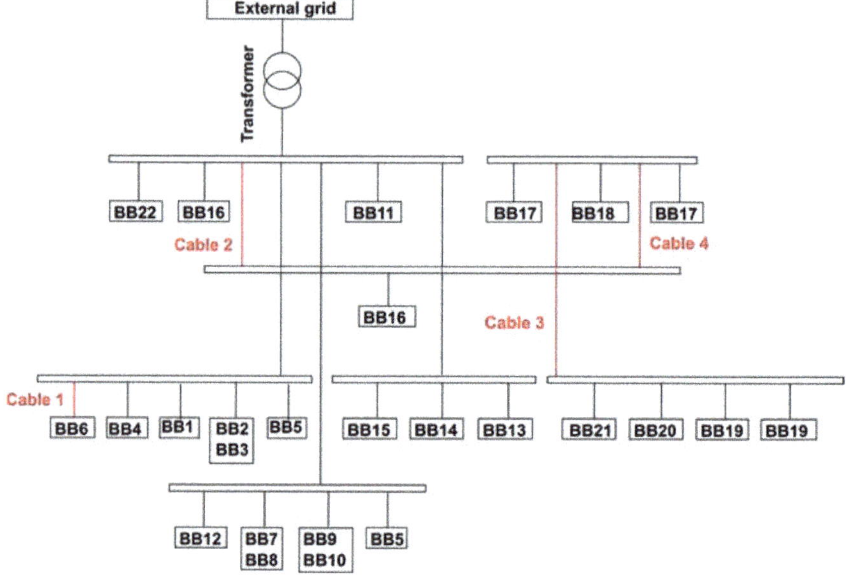

Figure 4. One-line diagram of the case study 400 V grid.

The stochastic driving patterns of the EVs were generated by means of an in-house, Python-based model. Table 2 summarizes the main inputs to the model, with the related references for the assumptions. These values were considered valid for the selected location. The number of cars per person referred to statistical data of a typical Stockholm neighborhood. To determine the density of inhabitants, it was reasonably assumed that two people could live in 48 m^2. The level of penetration of EVs was related to a future case, in line with the current trend of an increasing EV market share [34]. These three inputs were used by the model to determine the number of EVs per each BB.

Table 2. Inputs to the in-house EVs' driving profile model. Soc: state of charge.

Inputs	Value	Unit	Reference
Number of cars per person	0.377	#	[35]
Number of people per heated m^2	0.042	#	[12]
Share of EVs in the area	100	%	arbitrary
Travel distance boundaries	6–45	km	[36]
Speed boundaries	40–55	km/h	[36]
Power exchange	3.68	kVA	[37]
SoC boundaries	50–90	%	arbitrary
Minimum SoC for discharging	80	%	arbitrary
Number of power exchange stations	5	#	arbitrary
Number of trips per day	2	#	arbitrary

From further statistical data, it was possible to assume average driving kilometers and speeds. The state of charge (SoC) of an EV, which refers to the charge level of its battery, was constrained by charging and discharging limits. An EV had to be charged if its SoC was below 50%, while charging was prevented if its SoC was over 90%. This is mainly due to technical performance reasons of the battery. Discharging to the grid was enabled when an EV's SoC was over 80%. Since this represented a key parameter for the present study, a sensitivity analysis was performed for values between 70% and 90% of the minimum SoC for discharging. The number of charging stations for each BB and the number of trips per day referred to the context of a residential neighborhood. This was also in line with the initiative "Charge at Home," presented in Section 1. Thus, the two trips represented a way to work and a way back to home. The potential charging/discharging of an EV could happen after these two trips were completed each day.

Assumed initial hourly profiles for the share of cars starting a trip at a certain hour were taken as a further input to the model. These profiles were randomized using the parameters' ranges presented in Table 2.

Given all these inputs (Table 2), the in-house Python model performed three main steps:

- The total number of vehicles was allocated to each BB proportionally to the number of inhabitants and to the number of cars per person.
- The starting times for the cars' trips were randomized around an initial value defined per each hour of the day [36]. The "random" Python module was used for this purpose.
- The daily trips for each car were simulated by assigning, on a random basis ("random" module), the travel distances, the speed boundaries, and the initial SoC.

The main outputs of the model were:

- The number of cars per each station (or BB).
- The share of each cars' brand per each station (or BB).
- The starting and ending times of each trip per each car.
- The end-of-the-day SoC per each car.
- The available power hourly profile per each car per each station (or BB).

Finally, the EVs' available power profiles were combined with the overloading power profiles taken from the reference scenario. The objective was to estimate the potential of V2G to alleviate the grid overloadings caused by the installation of the distributed HPs. Thus, a peak power shaving strategy was implemented by using the electricity stored in the EV's batteries. This was done by overlapping the two hourly profiles along the selected critical day. In particular, a V2G discharging service was activated when the following combination of events occurred:

- A potential overloading was detected on an upstream cable.
- One or more EVs were parked with an SoC over 80% at a corresponding charging/discharging station.

The validity of the results was bound to the assumption that the EV's driving patterns did not present large differences from one day to the other. This was considered reasonable for a residential district, like Hammarby Sjöstad.

Since it was expected that the assumption for the minimum SoC for discharging played a relevant role, a sensitivity analysis was conducted for minimum SoC values equal to 70%, 75%, 80%, 85%, and 90%.

5. Results

Within the scope of the present paper, the results are presented and discussed in Figure 4 and in the remainder of this section.

Figure 4 highlights the cables that were found to be still critical within the reference scenario. This means that, despite the utilization of the thermal mass of the BBs as a TES device, not all the grid overloadings, caused by the HPs, could be compensated for [12]. This was mainly due to the constraints in terms of thermal mass capacity of each BB and indoor temperature comfort. The latter was assumed to 20 ± 0.5 °C. By introducing V2G technology, according to the models discussed in Section 4.2, the objective was to explore the technical potential of EVs' batteries to provide further capacity to the grid. In this way, it could be shown how V2G can help covering these remaining overloadings generated by the HPs' demand peak.

Concerning the V2G integration scenario, Table 3 shows the number of EVs that were assigned to each BB. This outcome was based on the parameters presented in Table 2, regarding the number of cars per person, the number of people per BB's heated square meters, and the share of EVs in the studied area.

Table 3. Number of EVs per each BB.

BB	Number of EVs
BB6	130
BB16	233
BB19	58
BB20	57
BB21	57

In terms of electricity discharging to the grid, the number of EVs reported in Table 3 shows the total availability of EVs when no constraints about trips and SoC were considered. However, within the V2G integration scenario, the EVs were firstly required to be parked at their BBs' stations after completing two daily trips (to and from work). Secondly, their SoC had to be above 80%. Based on these constraints, Figure 5 illustrates the EVs' availability when back from work, i.e., after the second daily trip. When looking at the number of EVs parked (top graph in Figure 5), it could be verified that the cars were driven back from work along the day, and, around 6 p.m., most of them were parked at home. However, since an SoC higher than 80% was also required, the actual, much lower, EVs' availability is shown in the bottom graph of Figure 5. For example, at the charging station of BB16, at 6 p.m., only 25% of the parked EVs could discharge electricity to the grid.

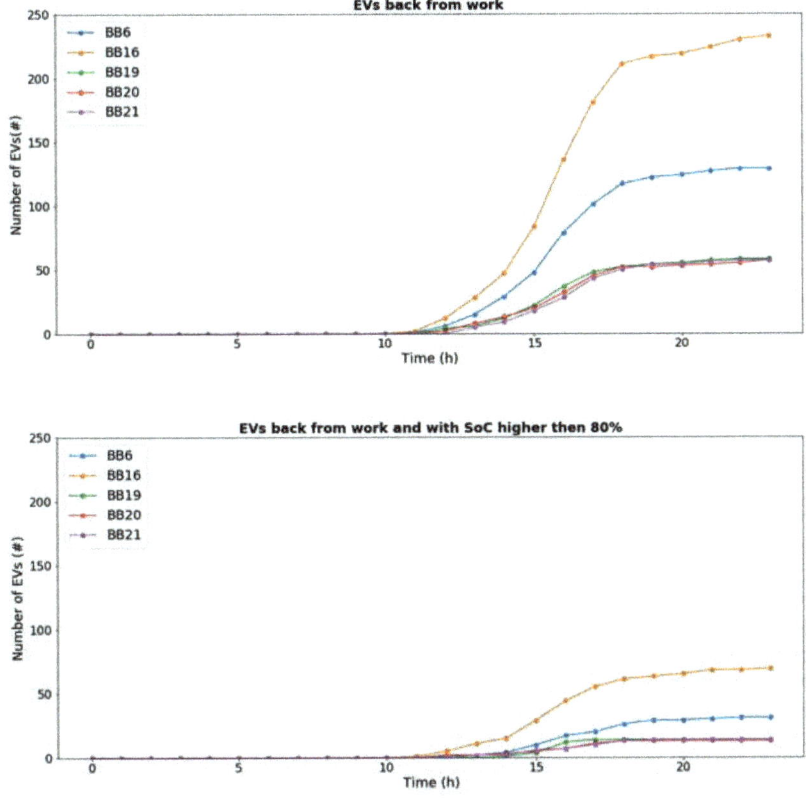

Figure 5. EVs' availability at each BB station during the selected critical day.

Figure 6 illustrates the impact of the EVs' availability for V2G on the grid's overloadings. In this figure, the reference scenario (without V2G integration) and the V2G integration scenario are compared over the selected critical day. The comparison is done by showing the percentage overloading performance for each critical cable. The loading performance below 100% is not shown in order to set the focus on the overloadings only.

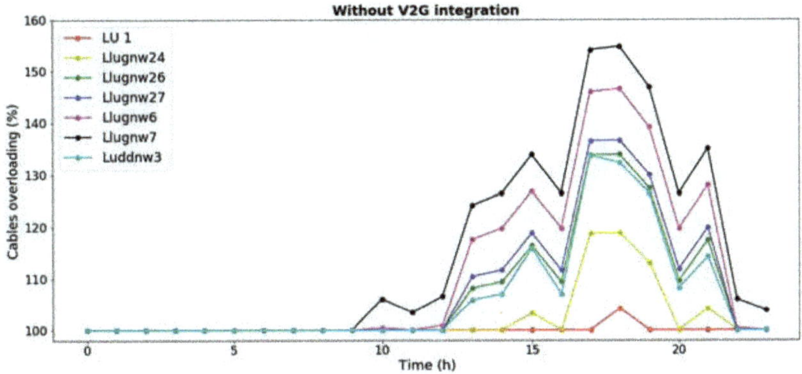

Figure 6. *Cont.*

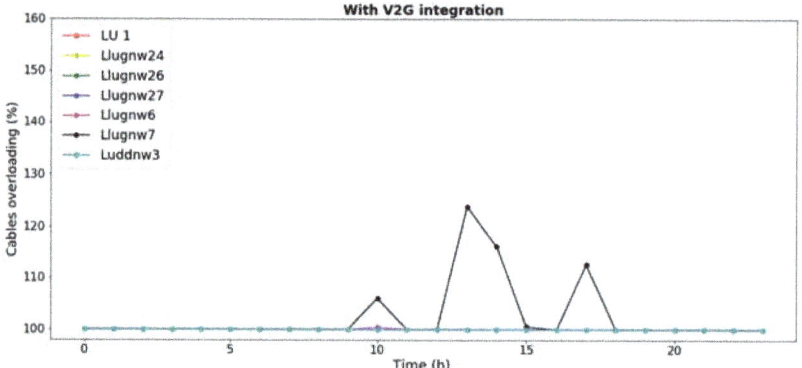

Figure 6. Cables overloading performance day before and after the integration of V2G.

As it can be noticed in Figure 6, the overloadings were concentrated during the late afternoon and during the evening. This is reasonable since the studied area is mainly a residential one, so people are at work during the day hours.

The integration of V2G helped to alleviate the overloadings by shaving the peak demand generated by the HPs. All the cables, except Llugnw6 and Llugnw7, fully benefited from this new system with no overloadings left. The other cables remained critical, especially before 6 p.m., when most of the cars were still away from the parking stations. This means that, with a 100% level of EV penetration in the neighborhood, the discharge of electricity from EVs' batteries could cover the amplitudes of the critical power peaks. The time matching between overloadings and the EVs' availability remained a challenge.

From the perspective of the main objective of the case study, Figure 7 shows the relation between the cables and each BB and, thus, each distributed HP unit. Considering that this analysis was related to a critical day (representative according to the assumption in Section 4.2), it can be concluded that the integration of V2G has the potential to fully enable the installation of HPs in BB6, BB19, BB20, and BB21. Concerning the other BBs, further measures should be implemented. For example, if a grid upgrade is to be either avoided or delayed, a different V2G operation logic should be tested. The implementation of smart control devices based on forecasting methods is a suggestion for future work.

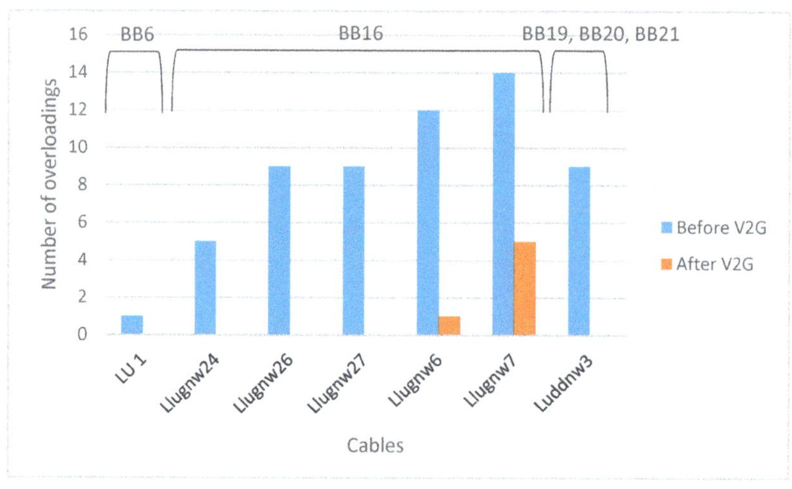

Figure 7. Cables overloadings over the selected critical day.

Finally, in order to assess the relevance of the assumption for the minimum SoC for discharging, a sensitivity analysis is presented for values equal to 70%, 75%, 80%, 85%, and 90%. It was found that the availability of cars with a minimum SoC of 70% allowed for the covering of all the overloadings during the selected critical day. However, this is expected to increase the following charging requirements, which should be further tested. On the opposite side, no cars were available when the minimum SoC was set to 90%, so none of the overloadings could be balanced. As an example, Figure 8 shows the results of this sensitivity analysis at 5 p.m. during the selected day for all the critical cables. At this specific hour, a minimum SoC of 75% already solved the overall criticality. However, the cable Llgnw7 was still overloaded by about 8% at 1 p.m.

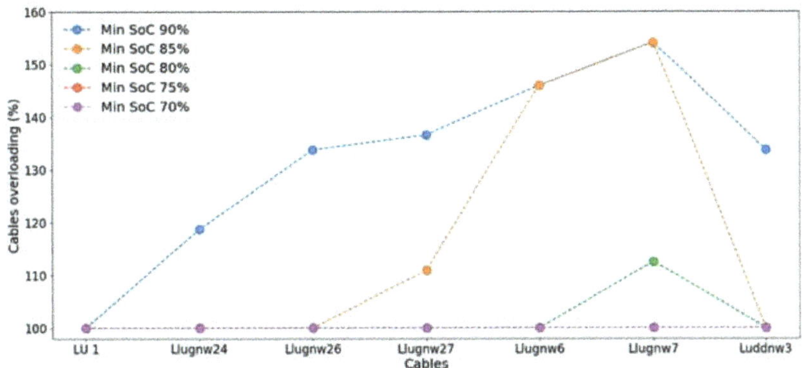

Figure 8. Cables' overloadings at 5 p.m. for different minimum SoC values for discharging.

As a more general conclusion, it can be stated that the approach presented in this paper can play a relevant role as a decision support tool for city planners, energy utilities, and engaged citizens. A district-level perspective linked to a sector-coupling approach (heating, electricity, and transport) can unlock the implementation of innovative technologies like active thermal mass and V2G.

6. Conclusions

In this paper, a district level perspective was applied on an integrated energy infrastructure problem where the electricity, heating, and transport sectors were interconnected.

Hammarby Sjöstad, a residential neighborhood in Stockholm, was selected as a relevant case. A previous study assessed that the plan of installing distributed domestic HPs will overload the local electricity distribution grid. The criticality of this situation can be improved, but not be solved, by using the thermal mass of the buildings as TES.

In the present work, V2G was presented as a new solution to further support the grid. This technology enables a bi-directional flow of electricity between the EVs' batteries and the grid. The aim was to explore the potential of V2G to perform peak power demand shaving by discharging electricity to the grid. The objective was to compensate for the overloadings caused by the HPs.

The technical performance simulation over a representative critical day demonstrated that, with a 100% penetration of EVs, the available power for discharging could cover all the overloadings' amplitudes. However, because of the time mismatch between the cars' availability and the need for balancing, only some cables could be completely relieved. This means that only the buildings connected to these cables could install residential HPs.

This conclusion is strongly dependent on the minimum SoC set for discharging. The study was conducted with a minimum SoC value of 80%. It was further shown that lowering this parameter to 70% could help solving the overall overloading problem. However, this is expected to have an impact on the following charging requirements, which should be further studied.

As future research, a model predictive control logic should be investigated in order to solve the time mismatch challenge. Furthermore, the validity of the results should be tested against potential daily EVs' driving patterns variations. A sensitivity analysis regarding the penetration level of EVs and their SoC boundaries for charging is also suggested.

Finally, the approach used in this study shows that city planners, energy utilities, and engaged citizen can benefit from taking a district-to-city level perspective on integrated energy systems.

Author Contributions: Conceptualization, M.A., and M.T.; methodology, M.A.; software, M.A. and M.T.; validation, M.A. and M.T.; formal analysis, M.A.; writing—original draft preparation, M.A.; writing—review and editing, M.T., M.A. and B.L.; visualization, M.A.; supervision, M.T. and B.L. All authors have read and agreed to the published version of the manuscript.

Funding: This research received no external funding.

Conflicts of Interest: The authors declare no conflict of interest.

References

1. Allerup, J. Sweden's Climate Act and Climate Policy Framework. Available online: http://www.swedishepa.se/Environmental-objectives-and-cooperation/Swedish-environmental-work/Work-areas/Climate/Climate-Act-and-Climate-policy-framework-/ (accessed on 15 October 2019).
2. IEA. *Sweden 2019 Review*; IEA: Paris, France, 2019.
3. Duveau, T.; Tewes, N. *Climate Scorecard Sweden*; WWF and Allianz: Frankfurt, Germany, 2009.
4. Moro, A.; Lonza, L. Electricity carbon intensity in European Member States: Impacts on GHG emissions of electric vehicles. *Transp. Res. Part. D Transp. Environ.* **2017**, *64*, 5–14. [CrossRef] [PubMed]
5. Electricity Hammarby Sjöstad 2.0. Available online: http://hammarbysjostad20.se/?lang=en (accessed on 12 December 2019).
6. Pasichnyi, O.; Wallin, J.; Kordas, O. Data-driven building archetypes for urban building energy modelling. *Energy* **2019**, *181*, 360–377. [CrossRef]
7. IEA. Nordic Energy Technology Perspectives 2016. In *Energy Technology Policy Diver*; IEA: Paris, France, 2016; p. 269.
8. Choi, W.; Wu, Y.; Han, D.; Gorman, J.; Palavicino, P.C.; Lee, W.; Sarlioglu, B. Reviews on grid-connected inverter, utility-scaled battery energy storage system, and vehicle-to-grid application-Challenges and opportunities. In Proceedings of the 2017 IEEE Transportation Electrification Conference, Chicago, IL, USA, 22–24 June 2017; pp. 203–210.
9. Yilmaz, M.; Krein, P.T. Review of the impact of vehicle-to-grid technologies on distribution systems and utility interfaces. *IEEE Trans. Power Electron.* **2013**, *28*, 5673–5689. [CrossRef]
10. Steward, D. *Critical Elements of Vehicle-to- Grid (V2G) Economics*; NREL: Golden, CO, USA, 2017.
11. IRENA. *Innovation Outlook: Smart Charging for Electric Vehicles*; IRENA: Abu Dhabi, UAE, 2019.
12. Arnaudo, M.; Topel, M.; Laumert, B. Techno-economic analysis of demand side flexibility to enable the integration of distributed heat pumps within a Swedish neighborhood. *Energy* **2020**, *195*, 117012. [CrossRef]
13. Saber, A.Y.; Venayagamoorthy, G.K. Plug-in vehicles and renewable energy sources for cost and emission reductions. *IEEE Trans. Ind. Electron.* **2011**, *58*, 1229–1238. [CrossRef]
14. Chukwu, U.C.; Mahajan, S.M. V2G electric power capacity estimation and ancillary service market evaluation. In Proceedings of the 2011 IEEE Power & Energy Society General Meeting, Detroit, MI, USA, 24–28 July 2011; pp. 1–8.
15. Wang, Z.; Wang, S. Grid power peak shaving and valley filling using vehicle-to-grid systems. *IEEE Trans. Power Deliv.* **2013**, *28*, 1822–1829. [CrossRef]
16. Kumar, A.; Bhalla, V.; Kumar, P. Impact of Plug-in Hybrid Electric Vehicles Integrated with Economic Unit Commitment of Power system. In Proceedings of the 2017 IEEE Transportation Electrification Conference (ITEC-India), Pune, India, 13–16 December 2017.
17. Shinde, P.; Swarup, K.S. Optimal Electric Vehicle charging schedule for demand side management. In Proceedings of the 2016 First International Conference on Sustainable Green Building Communities, Chennai, India, 18–20 December 2016.

18. Yang, Z.; Liao, Q.; Tang, F.; Peng, S.; Fang, F.; Xu, Y. Dispatch of EV Loads in Active Distribution Network Considering Energy Storage Characteristic. In Proceedings of the 2017 IEEE Conference on Energy Internet and Energy System Integration (EI2), Beijing, China, 26–28 November 2017.
19. Debnath, U.K.; Ahmad, I.; Habibi, D.; Saber, A.Y. Energy storage model with gridable vehicles for economic load dispatch in the smart grid. *Int. J. Electr. Power Energy Syst.* **2015**, *64*, 1017–1024. [CrossRef]
20. Ahmadian, A.; Sedghi, M.; Mohammadi-Ivatloo, B.; Elkamel, A.; Aliakbar Golkar, M.; Fowler, M. Cost-Benefit Analysis of V2G Implementation in Distribution Networks Considering PEVs Battery Degradation. *IEEE Trans. Sustain. Energy* **2018**, *9*, 961–970. [CrossRef]
21. Jain, P.; Jain, T. Development of V2G and G2V Power Profiles and Their Implications on Grid Under Varying Equilibrium of Aggregated Electric Vehicles. *Int. J. Emerg. Electr. Power Syst.* **2016**, *17*, 101–115. [CrossRef]
22. Freeman, G.M.; Drennen, T.E.; White, A.D. Can parked cars and carbon taxes create a profit? The economics of vehicle-to-grid energy storage for peak reduction. *Energy Policy* **2017**, *106*, 183–190. [CrossRef]
23. Gough, R.; Dickerson, C.; Rowley, P.; Walsh, C. Vehicle-to-grid feasibility: A techno-economic analysis of EV-based energy storage. *Appl. Energy* **2017**, *192*, 12–23. [CrossRef]
24. White, C.D.; Zhang, K.M. Using vehicle-to-grid technology for frequency regulation and peak-load reduction. *J. Power Sources* **2011**, *196*, 3972–3980. [CrossRef]
25. Mahmud, K.; Morsalin, S.; Kafle, Y.R.; Town, G.E. Improved peak shaving in grid-connected domestic power systems combining photovoltaic generation, battery storage, and V2G-capable electric vehicle. In Proceedings of the 2016 IEEE International Conference on Power Systems Technology, Wollongong, NSW, Australia, 28 September–1 October 2016; pp. 1–4.
26. Wang, Z.; Tang, Y.; Chen, X.; Men, X.; Cao, J.; Wang, H. Optimized daily dispatching strategy of building-integrated energy systems considering vehicle to grid technology and room temperature control. *Energies* **2018**, *11*, 1287. [CrossRef]
27. Lin, H.; Liu, Y.; Sun, Q.; Xiong, R.; Li, H.; Wennersten, R. The impact of electric vehicle penetration and charging patterns on the management of energy hub–A multi-agent system simulation. *Appl. Energy* **2018**, *230*, 189–206. [CrossRef]
28. Puerto, P. zerOBNL. Available online: https://github.com/IntegrCiTy/zerobnl (accessed on 26 October 2018).
29. Boverket Ska Din Byggnad ha en Energideklaration? Available online: https://www.boverket.se/sv/energideklaration/energideklaration/ (accessed on 27 June 2019).
30. Puerto, P.; Widl, E.; Page, J. ZerOBNL: A framework for distributed and reproducible co-simulation. In Proceedings of the 2019 7th Workshop on Modeling and Simulation of Cyber-Physical Energy System, Montreal, QC, Canada, 15 April 2019; pp. 1–6.
31. Pandapower. Available online: https://pandapower.readthedocs.io/en/v1.6.0/about.html (accessed on 11 October 2018).
32. Open Modelica Thermal Zone Reduced Order Models. Available online: https://build.openmodelica.org/Documentation/IBPSA.ThermalZones.ReducedOrder.RC.UsersGuide.html (accessed on 14 October 2019).
33. Patteeuw, D.; Helsen, L. Combined design and control optimization of residential heating systems in a smart-grid context. *Energy Build.* **2016**, *133*, 640–657. [CrossRef]
34. Bunsen, T.; Cazzola, P.; D'Amore, L.; Gorner, M.; Scheffer, S.; Schuitmaker, R.; Signollet, H.; Tattini, J.; Paoli, J.T.L. *Global EV Outlook 2019 to Electric Mobility*; OECD: Paris, France, 2019; p. 232.
35. Statistikmyndigheten SCB Statistikdatabasen. Available online: http://www.statistikdatabasen.scb.se/pxweb/sv/ssd/ (accessed on 1 February 2020).
36. Trafikverket Öppna Data från Trafikverket. Available online: https://www.trafikverket.se/tjanster/Oppna_data/ (accessed on 1 February 2020).
37. Liu, Z.; Wu, Q.; Nielsen, A.H.; Wang, Y. Day-ahead energy planning with 100% electric vehicle penetration in the nordic region by 2050. *Energies* **2014**, *7*, 1733–1749. [CrossRef]

Article

A Techno-Economic Centric Integrated Decision-Making Planning Approach for Optimal Assets Placement in Meshed Distribution Network Across the Load Growth

Syed Ali Abbas Kazmi [1,*], Usama Ameer Khan [1], Hafiz Waleed Ahmad [1], Sajid Ali [1] and Dong Ryeol Shin [2]

1. U.S.-Pakistan Center for Advanced Studies in Energy (USPCAS-E), National University of Sciences and Technology (NUST), H-12 Campus, Islamabad 44000, Pakistan; uxamaameer18@gmail.com (U.A.K.); 17eepwaleed@uspcase.nust.edu.pk (H.W.A.); sajidali75092@gmail.com (S.A.)
2. Department of Electrical and Computer Engineering, College of Information and Communication Engineering (CICE), Sungkyunkwan University (SKKU), Suwon 16419, Korea; drshin@skku.edu

* Correspondence: saakazmi@uspcase.nust.edu.pk; Tel.: +92-336-5727292

Received: 10 January 2020; Accepted: 3 March 2020; Published: 19 March 2020

Abstract: The modern distribution networks under the smart grid paradigm have been considered both interconnected and reliable. In grid modernization concepts, the optimal asset optimization across a certain planning horizon is of core importance. Modern planning problems are more inclined towards a feasible solution amongst conflicting criteria. In this paper, an integrated decision-making planning (IDMP) approach is proposed. The proposed methodology includes voltage stability assessment indices linked with loss minimization condition-based approach, and is integrated with different multi-criteria decision-making methodologies (MCDM), followed by unanimous decision making (UDM). The proposed IDMP approach aims at optimal assets sitting and sizing in a meshed distribution network to find a trade-off solution with various asset types across normal and load growth horizons. An initial evaluation is carried out with assets such as distributed generation (DG), photovoltaic (PV)-based renewable DG, and distributed static compensator (D-STATCOM) units. The solutions for various cases of asset optimization and respective alternatives focusing on technical only, economic only, and techno-economic objectives across the planning horizon have been evaluated. Later, various prominent MCDM methodologies are applied to find a trade-off solution across different cases and scenarios of assets optimization. Finally, UDM is applied to find trade-off solutions amongst various MCDM methodologies across normal and load growth levels. The proposed approach is carried out across a 33-bus meshed configured distribution network. Findings from the proposed IDMP approach are compared with available works reported in the literature. The numerical results achieved have validated the effectiveness of the proposed planning approach in terms of better performance and an effective trade-off solution across various asset types.

Keywords: distributed generation; distribution network; distribution network planning; distributed static compensator; losses minimizations; mesh distribution network; multi-criteria decision making; unanimous decision making; voltage stability assessment index

1. Introduction

The global load demand for electricity has increased significantly, pushing the distribution network (DN) to their operational limits results in issues i.e., voltage stability and system losses. Also, the distribution grid is more susceptible to technical, cost-economic, environmental, and social issues, especially from the perspective of meeting growing demand [1]. Conventionally, the traditional distribution grid paradigm was deterministically designed and planned to retain unidirectional power flow under radial topology, particularly considering simple protection schemes and easy control. Moreover, the traditional planning tools usually applied for distribution network planning problems (DNPP) might not remain feasible to mitigate the concerned issues by replication the existing infrastructure, which is certainly not a cost-effective solution [2]. The DNPP needs the support of various optimization tools of different genres aiming for futuristic scenarios.

Practically, DNPP aims realistically towards achieving a trade-off solution under multiple conflicting criteria subjected to various non-linear system constraints. The topology constraint in most of the DNPP studies has considered radial topology rather than interconnected configuration [3]. Similarly, distributed generation (DG) incorporation was neither considered in the planning stage nor assisted in the operational stage of the radial-structured distribution network (RDN) in the traditional grid paradigm. However, the addition of DG in DN has transformed the passive nature of the system into an active one and hence also transformed into an active distribution network (ADN) [4]. The RDN along with optimal DG placement (ODGP) can either remain in reconfigured configuration or be transformed into interconnected topologies i.e., loop DN (LDN) or mesh DN (MDN) on the basis of changing the state of normally open (NO) and tie-switches (TS). The interconnected arrangement is more suitable for densely inhabited urban centers and is feasible due to the cost-effectiveness of existing infrastructure employment [5,6].

In the recent literature studies, DNPP considering asset optimization has been considered as one of the core research dimensions to strengthen DN with various types of objectives subjected to constraints, with various methods applied at various system models. Plenty of techniques and methodologies of the different genres have been proposed for assets optimization studies (predominately DG) aiming at various single and multiple objectives (or criteria), which are usually conflicting in nature, under numerous constraints [7–10]. Among the main efforts to solve the aforementioned planning problems, the DN planners and utility operators consider optimal assets placement, dominated by DG units, in distribution mechanisms on the basis of size, location, quantity, capacity, type, and topology. The most sorted out solutions include cost-economic, technical, and environmental benefits, aiming at the achievement of trade-off solutions among multiple objectives [11]. The DNPP with DGs and associated assets have been considered a worthy solution, particularly enabling utilities to improve power quality and inducing deferral in DN up-gradation during load growth across the planning horizon, which usually spread across one year to several [12].

The methods addressing ODGP problems have been accredited to various objectives, primarily from the viewpoint of voltage (profile) maximization (VM) and system loss minimizations (LM). Moreover, technical advantages include DG penetration in DN, power quality (at utilities and consumers end), system stability, reliability, improved (bidirectional) power flows, and short-circuit-current (SCC) levels [7,8]. The other associated objectives concerned include the cost of active/reactive power losses, initial capital, operational, maintenance, and running cost. The environmentally feasible solutions with social acceptability concerning technology acceptance and consumer comfort are also amongst the addressed goals [9–11]. Besides DG, assets like reactive power compensation devices are also utilized for optimal operation of the DNPP, such as capacitors and flexible ac transmission system (FACTS) devices [6,12,13]. Furthermore, the application of distributed static compensators (D-STATCOMs) with DGs have been mostly reviewed in RDN [14]. In addition, normally open tie-switches (TS), normally closed sectionalize switches, concerned conductor replacements, and substation capacity enhancements are considered in asset optimization in DN. Furthermore, reconfiguration of a network to modified radial or to an interconnected topology has also been considered as a key component in the asset optimization of DN, from the viewpoint of DNPP [15,16].

In all of the above-mentioned works [6–16], the assets mostly used are DG, reactive power compensating devices, grid reinforcement with associated devices, and change of DN topology. The most significant asset optimization approaches aiming at optimal sitting and sizing under system constraints include classical techniques like analytical, deterministic, numerical, and exhaustive search.

The heuristic, meta-heuristic, artificial intelligence-based algorithms include nature-, society-, or population-inspired methodologies. However, these algorithms can result in local optima in various cases. This particular limitation is usually bridged with hybrid algorithms aiming at global optima. Besides that, multi-criteria (also known as multi-attribute) decision-making (MCDM) techniques are employed to sort out a trade-off solution among various concerned criteria/objectives of contradictory nature. Such methods can be priori optimized with assigning weights (subjectively or objectively) to each criterion (priori methods) or applied later (posteri methods) on a various number of solutions obtained from inner optimization. Besides that, commercial solvers are also employed for planning purposes such as the general algebraic modeling system (GAMS) [15,16].

From the perspective of DN, consideration of radiality constraint has dominated in most of the above-mentioned reviewed works, aiming at ODGP. However, the interconnected DN such as LDN and MDN are not as prevalent as their radial counterparts and need consideration from the viewpoint of planning [17]. LDN and MDN have been assessed from the perspective of various analytical/numerical and hybrid techniques from the perspective of various types of objectives such as loss minimization (LM) [18–20], voltage stabilization (VS) [19,20], DG penetration [19,20], reliability, and cost-related indices [20,21]. The LDN/MD-based infrastructure optimization has considered various assets such as the number of TS [22,23] and its influence on different load levels and evaluation across load growth [20,24,25]. Moreover, the replacement of TS with fault current limiter (FCL) [26], reinforcement versus looping/meshing [27,28], and optimal utilization of D-STATCOMs only in interconnected DN have also been considered [29]. In the recent works reported in [30,31], two different variants of an integrated planning approach incorporating improved voltage stability assessment indices (*VSAI*) along with loss minimization condition (*LMC*) have been employed for optimal asset optimization in MDN such as DG only and DG with D-STATCOM for VS, LM, and cost-related objectives under normal load.

The optimal planning of D-STATCOM is accredited with increasing penetration of renewable generation (REG)-based DGs, VS, LM, and minimizing associated cost objectives. Like most of the ODGP-based DNPP, D-STATCOM integration has mostly been considered radiality constraint [14]. It is also found that D-STATCOM has been utilized in mostly RDN for the achievement of core objectives such as system losses reduction, voltage curve improvement, and reduction of concerned costs. The work in [32] was aimed towards cost reduction along with the attainment of technical objectives. The D-STATCOM on the basis of asset placement on the same or different buses along with DGs or separately have been reported in [33–35], supported by relevant technical performance evaluations. The D-STATCOM placement on different load levels [36] and multiple asset sets (DG and D-STATCOM) on different buses [29–34] and the same buses [37] have been evaluated from various objectives. It is also important to mention that the works reported in [14,32–37] have mostly aimed at RDN, centered on a single branch (two buses) model and cannot encompass the core dynamic of LDN and MDN that is usually fed by more than on sending end.

From the viewpoint of MCDM, hybrid methodologies have been put forward to achieve multiple objectives or evaluation under various criteria. In the research works [38–41], the prominent MCDM methods employed are weighted sum method (WSM), weighted product method (WPM), technique for order preference by similarity to ideal solution (TOPSIS), preference ranking organization method for enrichment of evaluations (PROMETHEE), and RDN is reconfigured in terms of asset optimization to achieve objects such as active power losses, reliability, and average energy not served (AENS). Also, the heuristic and meta-heuristic methods in combination with MCDM have been utilized in various asset planning works to achieve a suitable solution. In [42], genetic algorithm (GA) and TOPSIS have been employed for optimal sitting and sizing of DG and remote terminal units (RTUs). In [43], DN is radially reconfigured with non-dominated GA-II (NSGA-II) and a combination of MCDM techniques to achieve an optimal solution with fewer energy losses, an optimum level of energy not served (ENS), and load balancing, respectively.

The particle swarm optimization (PSO) along with the analytical hierarchal process (AHP) in [44] have been utilized to achieve multi-objective solutions across technical, environmental, and economic-based criteria, with DGs in radially reconfigured DN. In [45], teaching–learning-based optimization is employed for multi-objectivity using penalty factors for a DG-only solution and an improved variant in [46] is used to achieve a solution under multiple assets (DGs and capacitors). The multi-objective, opposition-based, chaotic differential Equation-based method in [47] is used for techno-economic analysis of only DGs and to avoid premature convergence in the above-mentioned meta-heuristic methods. The research works in [48,49] have considered DG and renewable DG (REG) penetration along with various indices to offer a simple solution aiming at voltage stability and loss minimizations. The load growth has been briefly discussed for DGs only from the viewpoint of technical objectives considering MCDM in [50] and voltage stability index with MCDM-based methodology in LDN in [51,52].

As aforementioned, RDN was not planned to integrate DGs and nearly every DNPP with any sort of asset is aimed credibly towards achieving multiple conflicting objectives under any topology, abiding nonlinear system constraints. Thus, the modernization of DS in planning and operation with several DGs types and transformations to an ADN has become a noticeable research dimension. Hence, DN modernization with efficient asset optimization and interconnected topology can be considered as a prominent research dimension in the area of the smart distribution network (SDN) under the smart grid (SG) paradigm from the perspective of planning, scheduling, and operation, respectively. Moreover, the SDN under the SG paradigm is expected to be reliable from interconnected topology and multi-criteria attainment oriented with conflicting nature. The planning tools also need to be updated and evaluated across load growth considering multi-dimensional evaluation, since technically efficient solution might not be cost-effective. Hence, a composite asset planning problem with multi-criteria optimization needs research consideration, supported with multi-dimensional performance evaluation across load growth. Although reviewed works have partially addressed the aforementioned issues from various perspectives, bridging the limitations offered in reviewed literature is the motivating force for research and serves as the impetus of this paper.

In this paper, *VSAI* interrelated with *LMC*-based approach is integrated with various MCDM methodologies, followed by unanimous decision making (UDM), and is given the name integrated decision-making planning (IDMP). The MCDM methodologies employed in IDMP include WSM, WPM, TOPSIS, and PROMETHEE. The proposed IDMP approach aims at bridging the research gap in the reviewed literature by optimal asset optimizations in MDN for a trade-off solution amongst various alternatives across normal and load growth horizons. The 33-bus distribution system is configured to MDN as the precedence of ADN unlike their radial counterparts. The *VSAI* indices used in this approach are specifically designed and based on the multi-branch model and encompass the dynamics of an interconnected DN i.e., MDN, unlike radial counterparts based on a single branch model. The assets involved in IDMP have DG operating at various lagging power factors (LPF) contributing both active and reactive power, and renewable DG such as photovoltaic (PV) system contributes active power only and D-STATCOM units providing reactive power only. The approach provides alternatives across various axis such as technical only, economic only, and techno-economic objectives across the planning horizon. The MCDM methodologies provide a wide range of alternatives as solutions. In addition, unanimous decision making (UDM) across various MCDM methodologies in terms of their respective scores are offered. Moreover, the proposed IDMP approach can serve as a tool for future planning of interconnected ADN, particularly supporting planning engineers and researchers from the perspective of the SG paradigm. The main contributions of the proposed work are as follows.

(i) Integrated decision-making planning approach (IDMP) for optimal asset optimization.
(ii) Evaluation of the offered approach under various multiple assets (sitting and sizing) with LPF.
(iii) Evaluation of offered approach under various techno-economic performance metrics.
(iv) Detailed evaluation of alternatives across normal load and load growth horizons.
(v) Detailed evaluation of alternatives across four MCDM methodologies.

(vi) Offering a unanimous decision making (UDM) score as per rank of alternatives.
(vii) Numerical evaluations of the proposed approach on 33-Bus test DN.
(viii) Validations of achieved results with the findings reported in the available literature.

This paper is organized in the following sections. Section 2 offers the proposed IDMP approach along with concerned mathematical expressions. Section 3 offers a computation procedure for the IDMP approach, setup for simulations, and performance evaluation indicators. In Section 4, the attained numerical results regarding the effectiveness of the proposed approach is evaluated with multiple DG operating at various LPF, REG, and D-STATCOM sets, on the basis of optimal assets sitting and sizing perspective, evaluated under various performance metrics, demonstrated on 33-bus test MDN. The MCDM evaluations followed by UDM scores amongst various alternatives are presented in this section. The comparison of the proposed IDMP approach with existing research work is validated by comparison with existing works in Section 5. The paper concludes in Section 6.

2. Proposed Integrated Decision-Making Planning Approach

2.1. Voltage Stability Assessment Index_A (VSAI_A) for Mesh Distribution Network

The electrical equivalent MDN model in Figure 1 consists of three branches that represent DN feeders and two tie-line (TL) for in between linkage. The TS are closed to convert the DN into MDN. The voltages from sending end buses/nodes (n_{1b}, n_{3b} and n_{5b}) have been considered to exhibit the same magnitude and phase angle (δ), and is represented as one source node n_{1b}, respectively. The receiving end buses/nodes (n_{2b}, n_{4b}, and n_{6b}) are connected via two TB (with insignificant impedance) via respective TS. Also, the loads S_{2b}, S_{4b}, and S_{6b}, at n_{2b}, n_{4b}, and n_{6b} are considered as lumped load at bus/node m_{2b} with a voltage magnitude of V_{2b}, as shown in Figure 1, respectively.

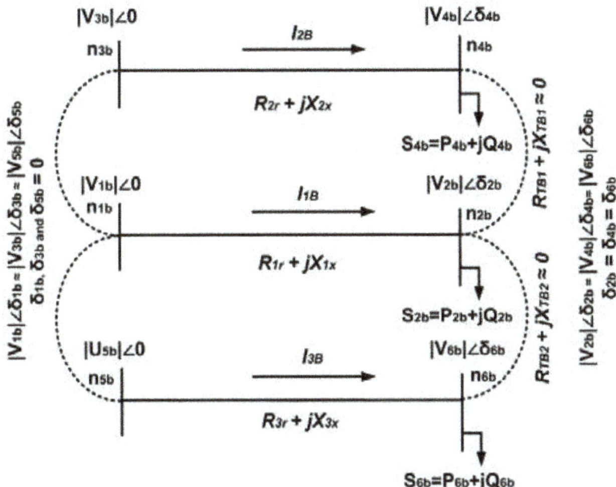

Figure 1. Electrical equivalent diagram of mesh distribution network [30,31].

In this paper, two of the VSAI indices formulated and reported in our previous publications [30,31] have been employed along with *LMC* aiming at pinpointing possible alternatives in terms of asset sitting as sizing with various decision variables. Later, various MCDM techniques are further applied for sorting out the best alternatives amongst available solutions. The VSAI reported in [30] is designated as VSAI_A and the other one reported in [31] is designated as VSAI_B, aiming at an optimal sitting of the asset. The procedure of *LMC* remains the same in both VSAIs as per their integrated planning approach. The VSAI_A along with feasible solution V_A are shown in Equations (1) and (2), where the variables are separately shown in Equations (3)–(6). The threshold value of VSAI_A value is between 0 (instability) and 1 (stable).

$$VSAI\_A = \sum_{i=1}^{n_l}\left(\frac{V_{sb}}{n}\right)^4 - \frac{4}{n}\sum_{i=1}^{n_l}\left(\frac{V_{sb}}{n}\right)^2\left[\left(\frac{A}{C}\right)+\left(\frac{B}{D}\right)\right] - \frac{4}{n^2}\left[\left(\frac{A}{C}\right)-\left(\frac{B}{D}\right)\right]^2 \geq 0 \quad (1)$$

$$V\_A = \frac{1}{\sqrt{2}}\sqrt{\left[\sum_{i=1}^{n_l}\left(\frac{V_{sb}}{n}\right)^2 - \frac{2}{n}\left\{\left(\frac{A}{C}\right)+\left(\frac{B}{D}\right)\right\}\right] + \sqrt{\sum_{i=1}^{m_l}\left(\frac{V_{sb}}{m}\right)^4 - \frac{4}{n}\sum_{i=1}^{n_l}\left(\frac{U_{sb}}{n}\right)^2\left[\left(\frac{A}{C}\right)+\left(\frac{B}{D}\right)\right] - \frac{4}{n^2}\left[\left(\frac{A}{C}\right)-\left(\frac{B}{D}\right)\right]^2}} \quad (2)$$

where

$$\begin{aligned} A = \; & P_{2B}\left[\prod_{i=1}^{n} R_{nr}\right]\left[1-\left(\frac{X_{1x}X_{2x}}{R_{1r}R_{2r}}+\frac{X_{1x}X_{3x}}{R_{1r}R_{3r}}+\frac{X_{2x}X_{3x}}{R_{2r}R_{3r}}\right)\right] \\ & + Q_{2B}\left[\prod_{i=1}^{n} nx\right]\left[\left(\frac{R_{1r}R_{2r}}{X_{1x}X_{2x}}+\frac{R_{1r}R_{3r}}{X_{1x}X_{3x}}+\frac{R_{2r}R_{3r}}{X_{2x}X_{3x}}\right)-1\right] \end{aligned} \quad (3)$$

$$\begin{aligned} B = \; & P_{2B}\left[\prod_{i=1}^{n} X_{nx}\right]\left[\left(\frac{R_{1r}R_{2r}}{X_{1x}X_{2x}}+\frac{R_{1r}R_{3r}}{X_{1x}X_{3x}}+\frac{R_{2r}R_{3r}}{X_{2x}X_{3x}}\right)-1\right] \\ & - Q_{2B}\left[\prod_{i=1}^{n} R_{nr}\right]\left[1-\left(\frac{X_{1x}X_{2x}}{R_{1r}R_{2r}}+\frac{X_{1x}X_{3x}}{R_{1r}R_{3r}}+\frac{X_{2x}X_{3x}}{R_{2r}R_{3r}}\right)\right] \end{aligned} \quad (4)$$

$$C = [abs\{\sum_{k\neq l}^{n} R_kR_l - \sum_{k\neq l}^{n} X_kX_l\} + \Delta Y];\ \Delta Y = 0.001 \quad (5)$$

$$D = abs(R_1X_2 + R_2X_1 + R_1X_3 + R_3X_1 + R_2X_3 + R_3X_2) = abs\left[\sum_{k\neq l}^{n} R_kX_l\right] \quad (6)$$

2.2. Voltage Stability Assessment Index_B (VSAI_B) for Mesh Distribution Network

The expression for *VSAI_B* is illustrated in Equation (7) and it is considered that unlike *VSAI_A* under normal conditions, the numerical value of *VSAI_B* is close to zero. During unstable conditions, the expression exceeds the numerical threshold of 1. The expression for *VSAI_B* and its feasible solution *V_B* are shown in Equations (7) and (8) with respective variables shown separately in Equations (9)–(10), respectively.

$$VSAI\_B = 4n^2\frac{\left[E\sum_{s=1}^{n_l}\left(\frac{V_{sb}}{n}\right)^2 + \left(\frac{F}{n}\right)^2\right]}{\sum_{s=1}^{n_l}\left(\frac{V_{sb}}{n}\right)^4} \leq 1 \quad (7)$$

$$V\_B = \frac{1}{\sqrt{2}}\sqrt{\left[\sum_{s=1}^{n_l}\left(\frac{V_{sb}}{n}\right)^2 - \frac{2E}{n}\right] + \sqrt{\sum_{s=1}^{n_l}\left(\frac{V_{sb}}{n}\right)^4 - \frac{4E}{n^2}\sum_{s=1}^{n_l}\left(\frac{V_{sb}}{n}\right)^2 - \frac{4F^2}{n^4}}} \quad (8)$$

where

$$E = [abs\{(P_{2b}R_{1r} + P_{4b}R_{2r} + P_{6b}R_{3r}) + (Q_{2b}X_{1x} + Q_{4b}X_{2x} + Q_{6b}X_{3x})\} + \Delta Y];\ \Delta Y = 0.001 \quad (9)$$

$$F = [abs\{(P_{2b}X_{1x} + P_{4b}X_{2x} + P_{6b}X_{3x}) - (Q_{2b}R_{1r} + Q_{4b}R_{2r} + Q_{6b}R_{3r})\} + \Delta Y];\ \Delta Y = 0.001 \quad (10)$$

2.3. Loss Minimization Condition (LMC) for Mesh Distribution Network

The *LMC* for *VSAI_A* and *VSAI_B* is the same at its optimal sizing of an asset at which the loop current across the tie-line is zero [30,31]. Figure 2 shows the electrical equivalent model of an equivalent MDN aiming at *LMC*. The loading at bus m_{2b} is considered at a normal load S_{2b}, fed by two TS ends (n_{4b} and n_{6b}) via tie-line currents (I_{TB1} and I_{TB2}), besides dedicated serving source (n_{1b}), respectively. The optimal sizing of assets that reduces I_{TB1} and I_{TB2} to zero indicates the optimal sizing of the respective asset. The *LMC* relations for base cases shown for apparent power (*LMC_S*) is shown in Equation (11). The optimal asset size at which I_{LP1} and I_{LP2} are zero represents the best case at which both active (P) and reactive (Q) power losses are minimized and indicated as the *LMC_P* and *LMC_Q* in Equation (12), respectively [30,31].

$$LMC\_S = \left[\left(I_{2B}' + I_{Lp1}\right)^2 Z_{2B} + \left(I_{1B}' + I_{Lp1} + I_{Lp2}\right)^2 Z_{1B} + \left(I_{3B}' + I_{Lp2}\right)^2 Z_{3B}\right] \quad (11)$$

$$LMC\_P + jLMC\_Q = \left[(I_{1B\_P}')^2 R_{1r} + (I_{2B\_P}')^2 R_{2r} + (I_{3B\_P}')^2 R_{3r}\right] + j\left[\left(I_{1B\_Q}'\right)^2 X_{1x} + \left(I_{2B\_Q}'\right)^2 X_{2x} + \left(I_{3B\_Q}'\right)^2 X_{3x}\right] \geq 0 \quad (12)$$

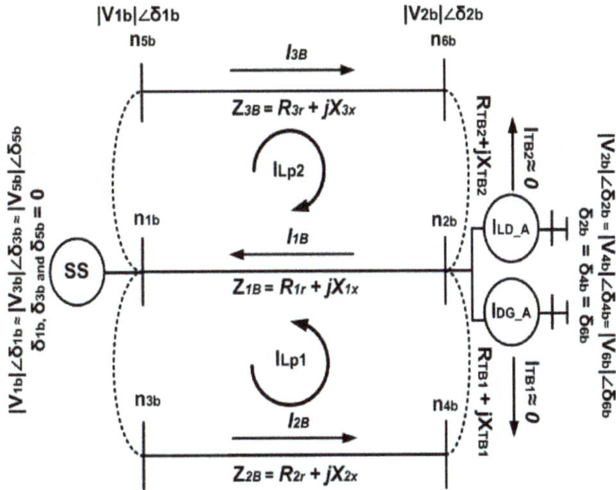

Figure 2. Equivalent mesh distribution network (MDN) model with tie and loop currents reduction aiming at loss minimization condition (*LMC*) [30,31].

2.4. Decision-Making (DM) Methodologies

The decision-making (DM) problems addressing multiple-attributes can be broadly classified among two types of classifications [38,41]. In the first type, referred to as the priori methods, weights are assigned (subjectively or objectively) to each criterion in case of predetermined solutions (also called alternatives). Such a method is also referred to as multi-attribute or multi-criteria decision-making. The second type refers to the posteri methods-based applications in which several solutions are initially obtained from inner optimization and, later, a best trade-off solution is acquired with any second-stage DM methodology. The methodologies related to this class have been utilized in the proposed IDMP approach. The generic decision matrix in MCDM is shown in Table 1.

Table 1. Generic decision matrix in multiple attribute decision-making (MCDM) methodologies.

Alternatives/Solutions	Weighted Attributes				
	$C_1{*}w_1$	$C_2{*}w_2$	$C_3{*}w_3$	…	$C_Y{*}w_Y$
A_1	S_{11}	S_{12}	S_{13}	…	S_{1y}
A_2	S_{21}	S_{22}	S_{23}	….	S_{2y}
A_3	S_{31}	S_{32}	S_{33}	…	S_{3y}
…	…	…	…	…	…
A_X	S_{X1}	S_{X2}	S_{X3}	…	S_{XY}

2.4.1. Weighted Sum Method (WSM)

The WSM is amongst the most used techniques for calculating the rank aiming at the achievement of the best solution among multiple solutions (also called alternatives) in terms of the highest score. For that purpose, the following equation finds the highest score solution as the optimum one considering m alternatives evaluated across n criteria.

$$S_{WSM} = \sum_{i}^{m} s_{ij} w_j \tag{13}$$

where i = 1,2, ... m, S_{WSM} indicates the weighted sum score, s_{ij} is the normalized score of i-th alternative/solution from the reference of j-th criterion and w_j is the weight associated with j-th criterion. Later on, the consequential cardinal scores for every alternative/solution can be utilized to rank or choose the best alternative. As aforementioned, the solution with the maximum score is considered as the best alternative amongst rest.

2.4.2. Weighted Product Method (WPM)

The WPM compares alternatives A_{kj} and A_{lj} across n criteria and the optimal solution is obtained by multiplication aiming at calculating ranks of alternatives rather than addition, as shown in WSM. The optimum solution in a pairwise comparison is the one that exhibits the highest score as shown in the Equation below.

$$S_{WPM} = \prod_{j=1}^{m} \left(\frac{A_{kj}}{A_{lj}} \right)^{w_j} = \prod_{j=1}^{m} \left(s_{ij} \right)^{w_j} \tag{14}$$

where i = 1,2, ... m, as previously, S_{ij} is the normalized score of the i-th alternative from the reference of j-th criterion and w_j is the weight associated with j-th criterion.

2.4.3. Technique for Order Preference by Similarity to Ideal Solution (TOPSIS)

After defining n criteria and m alternatives, the normalized decision matrix is established. The normalized value n_{ij} is calculated from Equation (15), where c_{ij} is the i-th criterion value for alternative Aj (j = 1 ... m and i = 1, ... , n).

$$n_{ij} = \frac{c_{ij}}{\sqrt{\sum_{j=1}^{m} c_{ij}^2}} \tag{15}$$

The normalized weighted values s_{ij} in the decision matrix are calculated as per Equation (16):

$$s_{ij} = n_{ij} w_j \tag{16}$$

The positive ideal A^+ and negative ideal solution A^- are derived as shown below, where I' and I'' are related to the benefit and cost criteria (positive and negative variables), as shown in Equation (17) as follows.

$$\begin{aligned} A^+ &= \{s_1^+, \ldots, s_1^+\} = \{(MAX_j s_{ij} | i \in I'), (MIN_j s_{ij} | i \in I'')\} \\ A- &= \{s_1^-, \ldots, s_1^-\} = \{(MIN_j s_{ij} | i \in I'), (MAX_j s_{ij} | i \in I'')\} \end{aligned} \tag{17}$$

From the n-dimensional Euclidean distance, $D_j{}^+$ is calculated in the given equation as the separation of every alternative from the ideal solution. The separation from the negative ideal solution is shown in a relationship indicated in Equation (18).

$$D_j^+ = \sqrt{\sum_{i=1}^{n} \left(sij - s_i^+ \right)^2} \; ; \; D_j^- = \sqrt{\sum_{i=1}^{n} \left(vij - v_i^- \right)^2} \tag{18}$$

The relative closeness C_j to the ideal solution of each alternative is calculated from Equation (19):

$$C_j = \frac{D_j^-}{\left(D_j^+ + D_j^- \right)} \tag{19}$$

After sorting the C_j values, the maximum value corresponds to the best solution to the problem.

2.4.4. Preference Ranking Organization Method for Enrichment of Evaluations-II (PROMETHEE-II)

The procedure of PROMETHEE II is indicated as follows.

Step 1: Normalize the decision matrix using the following Equation:

$$K_{ij} = \{L_{ij} - \min(L_{ij})\} / \{\max(L_{ij}) - \min(L_{ij})(i = 1,2,\ldots,\ n\ ,\ j = 1,2,\ldots,\ m) \tag{20}$$

where X_{ij} is the performance measure of i-th alternative with respect to j-th criterion. For non-beneficial criteria, Equation (20) can be rewritten as follows:

$$K_{ij} = \{max(L_{ij}) - (L_{ij})\} / \{\max(L_{ij}) - \min(L_{ij}) \tag{21}$$

Step 2: Calculate the evaluative differences of i-th alternative with respect to other alternatives. This step involves the calculation of differences in criteria values between different alternatives pairwise.

Step 3: Calculate the preference function, $P_j(i, i')$.

$$M_j(i, i') = 0\ if\ K_{ij} \leq K_{i'j} M_j(i, i') = (K_{ij} - K_{i'j})\ if\ K_{ij} \geq K_{i'j} \tag{22}$$

Step 4: Calculate the aggregated preference function taking into account the criteria weights. The aggregated preference function is shown in Equation (23) as follows, where w_j is the relative importance (weight) of j-th criterion is.

$$\pi(i, i') = \left[\sum_{j=1}^{m} w_j M_j(i, i')\right] / \sum_{j=1}^{m} w_j \tag{23}$$

Step 5: Determine the leaving and entering outranking flows, such as the leaving (or positive) flow for i-th alternative as indicated in Equation (23) and entering (or negative) flow for i-th alternative as shown in Equation (25), respectively; where n is the number of alternatives.

$$\varphi^{+}(i) = \frac{1}{n-1} \sum_{i'=1}^{n} \pi(i, i'), (i \neq i') \tag{24}$$

$$\varphi\text{-}(i) = \frac{1}{n-1} \sum_{i'=1}^{n} \pi(i', i), (i \neq i') \tag{25}$$

Step 6: Calculate the net outranking flow for each alternative as per Equation (26).

$$\varphi(i) = \varphi^{+}(i) - \varphi\text{-}(i) \tag{26}$$

Step 7: Determine the ranking of all the considered alternatives depending on the values of $\varphi(i)$. When the value of $\varphi(i)$ is higher, the alternative is preferred in terms of the best solution.

2.5. *Unanimous Decision Making (UDM) and Unanimous Decision Making Score (UDS)*

The trade-off solution via aforementioned MCDM techniques results in multiple best solutions across various cases of assets placement with respective scenarios. To find a unanimous best solution amongst the abovementioned MCDM techniques, unanimous decision making (UDM) is applied. Initially, the achieved rank is arranged as per the highest to the lowest best solution and is designated by A_R. Similarly, each arranged rank is given a score designated by A_S, such as the highest rank for example 1 will have the highest score like N, as shown in Table 2.

Table 2. Initial rank and score allocation for unanimous decision making (UDM).

Alternatives Rank (A_R) (Highest to Lowest)	Alternatives Score (A_S) (Highest to Lowest)
$A_{1R} = 1$	N
$A_{2R} = 2$	N-1
$A_{3R} = 3$	N-2
…	…
$A_{XR} = N$	1

The unanimous decision-making score (UDS) can be found via the following relationship as shown in Equation (27), across the finding of each MCDM technique. Since four techniques are considered, the solution will run across these techniques across all alternatives, terminating at a UDS. This UDS will determine the highest rank on the basis of the highest numerical value and is designated here as a unanimous decision-making rank (UDR). In this case, two UDR scores are equal, and the one with at least one of the highest alternatives will be given preference over others.

$$UDS = \sum_{MCDM=1}^{n=4} \left(A_R \times \frac{A_S}{A_R} \right) = \sum_{MCDM=1}^{n=4} (A_S) \tag{27}$$

3. Proposed Integrated Decision-Making Planning (IDMP) Approach, Computation Procedure, Constraints, Simulation Setup, and Performance Evaluation Indicators

3.1. Proposed Integrated Decision Making Planning (IDMP) Approach

It is one of the core responsibilities of power utility companies to supply sustained voltage levels with a feasible level of power quality to consumers via DN at each branch. Usually, the core aim is to achieve a win–win situation in favor of both utilities and consumers across a certain planning horizon. The load difference among distribution branches across various planning horizons can increase system losses in a DN. The actual planning problem is also aimed towards the attainment of objectives rather than a single one. The objectives usually aimed at a DN are inclined towards technical and cost-economic ones; whereas the cost of planning and operation is a vital factor in the distribution of power to respective load centers.

The flow chart of the proposed IDMP approach is illustrated in Figure 3. The reason for utilizing *VSAIs* in [30,31] for MDN for the proposed planning approach is that the planning problems associated with interconnected networks such as LDN and MDN do not have unique solutions like RDN. Also, these *VSAIs* are particularly designed to encompass all the prerequisites of an actual MDN, where there are usually more sending ends supplying the load. The *LMC* in both [30,31] is the same, aiming at the reduction of tie-line current and maintaining equal voltage across respective tie-switches, which are closed to transform an RDN into MDN. The consideration of MDN is also considered valid since it closely corresponds to the future ADN that is considered both interconnected and reliable. The main aim of each MCDM strategy utilized late, is to find the feasible planning solution, capable of achieving maximum relevant goals. Due to different solutions via each MCDM methodologies, a unanimous decision to follow becomes a necessity as a tool for following a solution that is best across technical, cost economic, and overall dimensions.

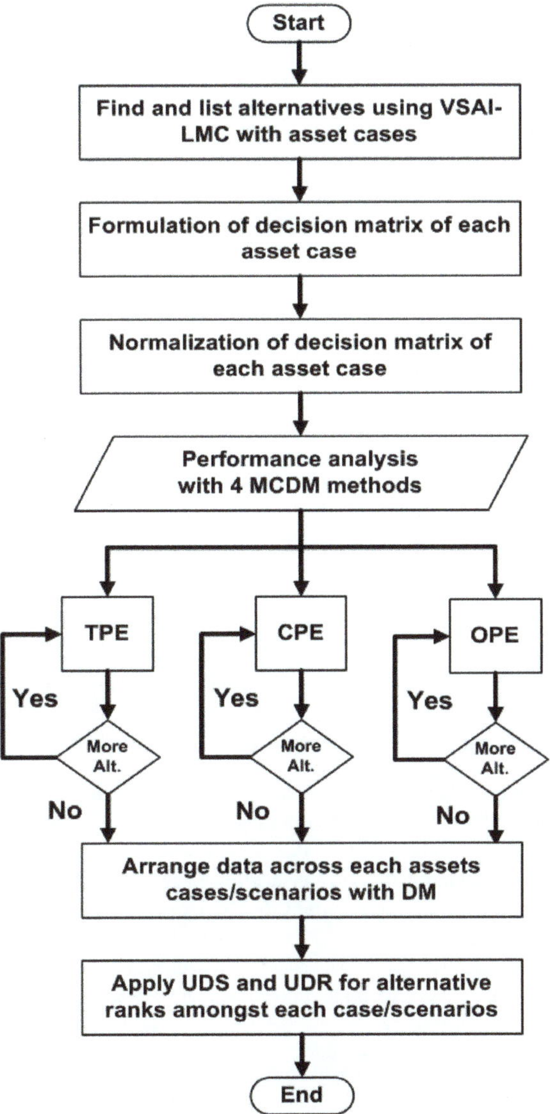

Figure 3. Flow chart of proposed integrated decision-making planning (IDMP) approach.

In the proposed IDMP approach, initially in stage 1, two different *VSAIs* (*VSAI_A* and *VSAI_B*) with respective *LMC* aims at optimal sitting and sizing of assets in terms of finding suitable alternatives, as per Equations (1)–(11), respectively. The assumptions for both *VSAI_A* and *VSAI_B* are the same and can be found for normal load only in [30,31]. Later in stage 2, four different MCDM approaches are individually applied on the attainment of best alternative amongst technical criteria only, cost-economic criteria only, and overall (techno-economic) criteria across various cases of asset optimization, as per Equations (12)–(24).

In Figure 3, it is shown that the performance is assessed on the basis of technical performance evaluation (TPE), cost-economic performance evaluation (CPE), and overall combined (techno-economic) performance evaluation (OPE). Finally, in stage 3, UDM is applied and the best solution amongst multiple MCDM is sorted out on the basis of their respective UDM scores (UDS) of each case as per Section 2.4, particularly Equation (25). The achieved UDS will define new UDR of alternatives across multiple MCDM methodologies. The weights throughout each MCDM methods have been considered as equal or unbiased weighting since such weight is mostly utilized in most of the planning problems involving conflicting criteria. All the predetermined alternatives and the trade-off final solution achieved are subjected to practical system constraints.

3.2. Computation Procedure of Proposed IDMP Approach for Alternatives Selection and Case Studies

The overall computation method for *VSAI_A* and *VSAI_B* reported in [30,31] is essentially the same with a little difference. The *LMC* approach integrated with each VSAI is the same. The decision variables (DV) for optimal asset placement are based on type, size, and the number of assets. In [30], two variants of the planning approach were based on *VSAI_A* and *LMC*, whereas in [31], a single approach was presented based on *VSAI_B* and *LMC*. The assets are considered on the basis of numbers achieved in the previous publications [30,31]. The alternatives with DG were evaluated across normal load levels in [30], whereas the alternatives with DG only and asset sets (REG + D-STATCOM) were evaluated across normal load levels [31]. However, the abovementioned assets (DG only and REG + D-STATCOM) were evaluated across load growth levels for this study. The alternatives were achieved as follows:

1. Alternate 1 (A1): 1×DG [30] or 1×asset set (REG + D-STATCOM) with *VSAI_A* and *LMC*.
2. Alternate 2 (A2): 1×DG or 1×asset set (REG + D-STATCOM) with *VSAI_B* and *LMC* [31].
3. Alternate 3 (A3): 2×DG [30] or 2×asset sets (REG + D-STATCOM) with *VSAI_A* and *LMC*.
4. Alternate 4 (A4): 2×DG or 2×asset sets (REG + D-STATCOM) with *VSAI_B* and *LMC* [31].
5. Alternate 5 (A5): 3×DG [30] or 3×asset sets (REG + D-STATCOM) with *VSAI_A* and *LMC*.
6. Alternate 6 (A6): 3×DG [30] or 3×asset sets (REG + D-STATCOM) with *VSAI_A* and *LMC*.
7. Alternate 7 (A7): 3×DG or 3×asset sets (REG + D-STATCOM) with *VSAI_B* and *LMC* [31].

All of the above seven alternatives have been evaluated in four cases across four MCDM methodologies under normal load (NL), load growth (LG) across five years, and optimal load growth (OLG) across five years, respectively. In NL, the current load is considered, and all the cases are evaluated. In LG, a 7.5% increment in load per annum is considered across five years, and asset sizing obtained during NL is retained as constant. In OLG, optimal asset sizing is considered across incremented load across five years. The OLG corresponds to the reinforcement required to maintain a solution after a planning horizon is over. The cases for evaluation with respective scenarios are TPE across technical criteria, CPE across cost-related criteria, and OPE across combined techno-economic criteria, respectively. The nomenclature of considered cases with respective scenarios for overall evaluation in this paper is presented as follows.

Case 1: DGs only assets placements in MDN operating at 0.90 lagging power factor (LPF).
Case 2: DGs only assets placements in MDN operating at 0.85 LPF.
Case 3: Asset set (REG + D-STATCOM) placements in MDN equal to 0.90 LPF.
Case 4: Asset set (REG + D-STATCOM) placements in MDN equal to 0.85 LPF.

In all of the above-mentioned four cases with respective designations, each case (C#) has been evaluated across the following scenarios of MCDM evaluations under various load levels as presented below.

Scenario 1 (NL):

Case 1 (C1_NL): TPE, CPE and OPE with WSM under NL.
Case 2 (C2_NL): TPE, CPE and OPE with WPM under NL.
Case 3 (C3_NL): TPE, CPE and OPE with TOPSIS under NL.

Case 4 (C4_NL): TPE, CPE and OPE with PROMETHEE under NL.

Scenario 2 (LG):

Case 1 (C1_LG): TPE, CPE and OPE with WSM under LG.
Case 2 (C2_LG): TPE, CPE and OPE with WPM under LG.
Case 3 (C3_LG): TPE, CPE and OPE with TOPSIS under LG.
Case 4 (C4_LG): TPE, CPE and OPE with PROMETHEE under LG.

Scenario 3 (OLG):

Case 1 (C1_OLG): TPE, CPE and OPE with WSM under OLG.
Case 2 (C2_OLG): TPE, CPE and OPE with WPM under OLG.
Case 3 (C3_OLG): TPE, CPE, and OPE with TOPSIS under OLG.
Case 4 (C4_OLG): TPE, CPE and OPE with PROMETHEE under OLG.

Later, UDM is applied across all cases under all the above-mentioned cases with respective scenarios for a unanimous solution via UDS and the result attained in terms of UDR. The highest UDS value refers to the best solution with the highest UDR.

3.3. Constraints Considered in Simulations

The following main constraints [30,31] have been considered in this study. It was ensured that the simulations do not result in a solution that results in reverse power flow towards substation. It is considered that active and reactive power contribution from substation (P_{SS}, Q_{SS}) and DG or other asset units (P_{DG}, Q_{DG}) must have a balance that is equal to active and reactive power load consumption (P_{LD}, Q_{LD}) along with associated active and reactive power losses (P_{Loss}, Q_{Loss}) in MDN.

$$(P_{SS} + Q_{SS}) + (P_{DG} + Q_{DG}) = (P_{LD} + Q_{LD}) + (P_{Loss} + Q_{Loss}) \tag{28}$$

The magnitude of the voltage at each node/bus "n" in MDN must not exceed the specified limit of 0.95 P.U to 1.05 P.U.

$$0.95 \le Vn \le 1.05;\ n = 1, 2, 3, 4 \ldots\ m \tag{29}$$

The LPF of DG is kept within limits considering an allowable variation of ±3%.

$$PF_{DG,\ i,min} \le PF_{DG,i} \le\ PF_{DG,i,max} \tag{30}$$

3.4. Simulation Setup

The proposed IDMP approach is tested on the 33-bus test distribution network (TDN), as displayed in Figure 4. The P_{LD} and Q_{LD} in the 33-bus TDN account for 3715 KW and 2300 KVAR, whereas P_{Loss} and Q_{Loss} account for 210.9 KW and 143.02 KVAR during a base case under normal load, respectively. In the case of load growth, The P and Q loads are 5333.363 KW and 3301.95 KVAR, respectively. The P and Q losses during load growth account for 450.65 KW and 305.17 KVAR, respectively.

The test MDN consists of four branches and five TSs. The 33-bus TDN is converted into a multiple-loop configured MDN by closing TS4 and TS5 (highlighted in green solid line) and results in two loop currents (ILp1 and ILp2) across two TB, respectively. The load or power flow analysis regarding the 33-bus TDN is obtained in terms of numerical values from equivalent models and has been implemented on MATLAB R2018a. The test setup is developed in SIMULINK and numerical values are called in m-files where the proposed approach is evaluated into achieved results. Initially, the assets are placed on designated locations given by *VSAIs* in [30,31]. The base case model is made in SIMULINK and values are called in m-file that indicate the weakest nodes as shown in [30,31]. Later, the numerical values were obtained from simulation setup in SIMULINK and are run until the condition where loop currents across TB are near zero and voltages across the respective nodes are equal with the optimal sizing of assets considering termination criteria of 1%. Finally, on termination, the achieved values are called in a program made of m-files (MATLAB 2018a), where the proposed IDMP approach is evaluated with various matrices and is represented in the following Section 3.5.

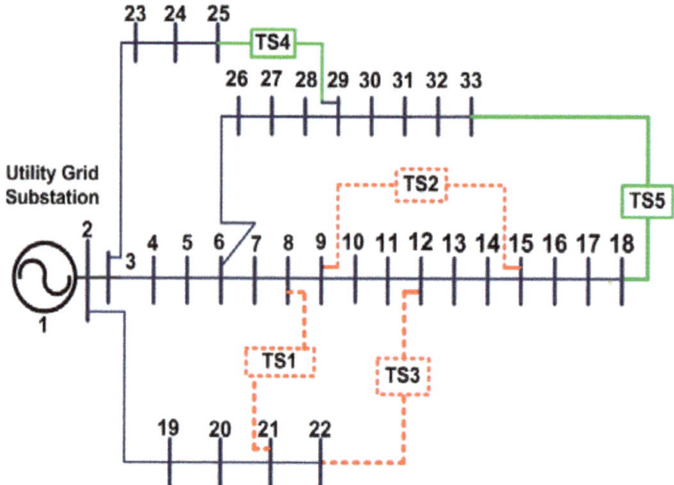

Figure 4. 33-bus (meshed configured) test distribution system.

The proposed IDMP approach is evaluated across various load and generation levels in terms of snapshot analysis. Although the IDMP approach seems dynamic, the snapshot analysis considering NL, LG, and OLG is evaluated in terms of steady-state analysis. The approach simplifies the requirement of dynamic analysis and gives a big picture in terms of steady-state analysis.

The asset sitting and sizing cases are evaluated across NL, where optimal sizing of assets is achieved in terms of performance evaluations. In LG, the generation from NL cases is kept constant and an increase in load after five years is considered for evaluation of the assets performance across the planning horizon. In OLG, the assets reinforcements required for optimal performance after planning horizon are evaluated.

3.5. Performance Evaluation Indicators (PEI)

The technical and cost-economic indices for performance evaluation are illustrated in Tables 3 and 4 and all relations have been taken from our previous publications in [30,31].

Table 3. Technical indices for performance evaluation (designated as technical performance evaluation (TPE)).

S#	Performance Indices	Performance Indices Relationships	Units	Objective
1	Active Power Loss (P_{Loss})	$PLMC' = \min \sum_{i=1}^{m_l-1} \mathrm{P_{Loss}}^{TDS} + \sum P_{TB}$	KW	Decrease
2	Reactive Power Loss (Q_{Loss})	$QLMC' = \min \sum_{i=1}^{m_l-1} \mathrm{Q_{Loss}}^{TDS} + \sum Q_{TB}$	KVAR	Decrease
3	Active Power Loss Minimization (PLM)	$PLM = \left[\frac{P\_L_{No\_DG} - P\_L_{M\_DG}}{P\_L_{No\_DG}}\right] \times 100$	%	Increase
4	Reactive Power Loss Minimization (QLM)	$QLM = \left[\frac{Q\_L_{No\_DG} - Q\_L_{M\_DG}}{Q\_L_{No\_DG}}\right] \times 100$	%	Increase
5	DG Penetration by percentage ($DGPP$)	$PDG = \left(\sum_{a=1}^{M} P_{DG} / \sum_{b=1}^{N} P_{LD}\right) \times 100$	%	Increase
6	Capacity Release of Active Power from Substation (P_{SSR})	$P_{SSR} = P_{SS} - P_{DG} \geq 0$	KW	Decrease
7	Capacity Release of Reactive Power from Substation (Q_{SSR})	$Q_{SSR} = Q_{SS} - Q_{DG} \geq 0$	KVAR	Decrease
8	Voltage Level	V = 1.0 P.U	P.U	Decrease

Table 4. Cost-economic indices for performance evaluation (designated as cost-economic performance evaluation (CPE)).

S#:	Performance Indices/Ref	Performance Indices Relationships	Units	Objective
1	Cost of active power loss (PLC)	$PLC = [P\_L \times E_U \times T_Y\ (8760\ hrs)]$	M$	Decrease
2	Active power loss saving (PLS)	$PLS = \frac{PLC_{No\_DG} - PLC_{M\_DG}}{PLC_{No\_DG}} \times 100$	M$	Increase
3	Cost of DG for P_{DG} ($CPDG$)	$C(P_{DG}) = a \times P_{DG}^2 + b \times P_{DG} + c$ where: a = 0, b = 20, c = 0.25	$/MWh	Decrease
4	Cost of DG for Q_{DG} ($CQDG$)	$C(Q_{DG}) = \left[C(S_{DG\_M}) - C\left(\sqrt{(S_{DG\_M}^2 - P_{DG}^2)}\right)\right] \times k$ where: $S_{DG_M} = \frac{P_{DG_M}}{\cos\theta} = \frac{1.1 \times P_{DG}}{\cos\theta}; k = 0.5 - 1$	$/MVArh	Decrease
5	Annual Investment Cost (AIC) $AIC\left(Million \frac{US\$}{Year}\right)$	$\sum_{k=1}^{M_{DG}} AF_C \times CU_C \times DGC_{max};$ Where : $AF_c = \frac{\left(\frac{Ct}{100}\right)\left(1+\frac{Ct}{100}\right)^T}{\left(1+\frac{Ct}{100}\right)^T - 1}$	Millions USD (M$)	Decrease
6	Annual Cost of D-STATCOM (ACD)	$I_C = \left[\frac{(1+C)^{nD} \times C}{(1+C)^{nD} - C}\right];$ where: I_C = 50$/KVAR; C = Rate of Assets return = 0.1; nD (in years) = 5	Millions USD (M$)	Decrease

4. Results and Discussions

The IDMP approach is applied in three stages as aforementioned in Sections 3.1 and 3.2. The first stage is employed for the layout (sitting and sizing) of numerous assets such as DG and D-STATCOM units in the MDN. The proposed integrated approach consists of two parts; *VSAIs* [30,31] are applied for potential assets (DG and REG + DSTATCOM) locations for sitting and *LMC* for optimal asset sizing.

In total, seven alternatives were shortlisted encapsulating four cases of assets sitting and sizing, across NL, LG, and OLG, respectively. Case 1 covers DGs operating at 0.90 LPF, case 2 covers DG operating at 0.85 LPF, and case 3 covers renewable DG such as PV system that contributes active power (P) only and reactive component (Q) comes from D-STATCOM. The contribution from the set of these two assets i.e., the P and Q contributions, is equal to 0.90 PF. In case 4, P and Q contributions are equal to that of one DG contributing at 0.85 LPF. Cases 1–2 are different than cases 3–4 in such a manner that DG only cases can be subjected to reactive power instability whereas in later cases, the power sources are decoupled. So, a comparative analysis is justified in terms of performance analysis.

In the second stage, four MCDM methodologies are applied to find out the best solution amongst the sorted alternatives. In the third stage, unanimous decision making (UDM) is applied to find out a common best solution in the achieved solutions that may vary on the basis of MCDM techniques.

The proposed IDMP approach is evaluated across technical, cost-economic, and combined techno-economic criteria of conflicting nature. Since the cost-related criteria may differ from various asset types, for the sake of composite evaluation, separated P and Q injections are considered.

4.1. Case 1 under Normal Load (C1_NL): DGs Only Placements in MDN Operating at 0.90 LPF

4.1.1. Initial Evaluation of Alternatives in Case 1 under Normal Load (C1_NL)

The initial evaluation of C1_NL for each alternative is shown in terms of TPE and CPE are shown in Table 5. The numerical values refer to evaluated indices values as potential criteria results obtained for seven alternatives referring to DG only asset placement operating at 0.90 LPF under NL. The reason being such a PF is favored by utilities.

Table 5. Techno-economic evaluation analysis in case 1 (C1_NL) for 33-bus MDN.

S#		Technical Parameters Evaluations (TPE)							Cost (Economics Related) Parameters Evaluations (CPE)					
Case (No.) / Alt. (No).	DG Size (KVA) @ Bus Loc.	P_{Loss} (KW)	Q_{Loss} (KVAR)	*PLM* (%)	*QLM* (%)	*DGPP* (%)	V_{Min} (P.U)	P_{SSR} + j Q_{SSR} (KW) + j(KVAR)	*PLC* (M$)	*PLS* (M$)	*CPDG* ($/MWh)	*CQDG* ($/MVArh)	*AIC* (M$)	*ACD* (M$)
C1/A1	DG1: 2013@15	125	89	40.76	37.76	46.07	0.9725	2028 + j 1511.5	0.0657	0.0452	36.47	3.6269	0.36374	0
C1/A2	DG1: 2750@30	38.3	28.933	81.85	79.76	63	0.9764	1278.3 + j 1130.23	0.02013	0.09077	49.75	4.9527	0.4969	0
C1/A3	DG1: 971@15 DG2: 1783@30	54.7	37.5	77	73.78	63	0.9750	1269.4 + j 1125.05	0.0174	0.0822	49.822	4.96	0.49763	0
C1/A4	DG1: 2357@30 DG2: 540@25	32.99	25.491	84.37	82.17	66.303	0.9769	1140.7 + j 1062.72	0.01261	0.09829	52.396	5.2141	0.5235	0
C1/A5	DG1: 832.6@15 DG2: 1602@30 DG3: 745.1@7	30.85	23.29	85.38	83.71	72.77	0.9904	884.12 + j 937.29	0.0162	0.0947	57.485	5.8832	0.5750	0
C1/A6	DG1: 894.6@15 DG2: 1386@30 DG3: 822.6@25	33.2	23.94	84.27	83.26	71.02	0.9891	955.32 + j 971.2865	0.0174	0.0935	56.107	5.5889	0.5607	0
C1/A7	DG1: 1957@30 DG2: 500@25 DG3: 760@8	18.87	13.327	91.06	90.68	73.625	0.9857	838.57 + j 911.069	0.00992	0.1010	58.156	5.7938	0.5813	0

Table 6. Order of the ranks across TPE, CPE, and OPE in C1_NL for 33-bus MDN.

Evaluations	TPE (C1_NL)						CPE (C1_NL)						OPE (C1_NL)					
C1_NL / Alt (#)	WSM	WPM	TOPSIS	PROMETHEE	UDS	UDR	WSM	WPM	TOPSIS	PROMETHEE	UDS	UDR	WSM	WPM	TOPSIS	PROMETHEE	UDS	UDR
A1	7	7	7	7	**4**	**7**	2	7	7	1	**15**	**4**	7	7	7	7	**4**	**7**
A2	5	5	5	5	**12**	**5**	3	3	2	2	**22**	**2**	5	5	5	5	**12**	**5**
A3	6	6	6	6	**8**	**6**	7	6	6	4	**9**	**7**	6	6	6	6	**8**	**6**
A4	3	3	2	4	**20**	**3**	4	2	1	3	**22**	**1**	3	3	2	4	**20**	**3**
A5	2	2	4	3	**23**	**2**	6	5	5	7	**9**	**6**	2	2	4	3	**21**	**2**
A6	4	4	3	2	**19**	**4**	5	4	4	5	**14**	**5**	4	4	3	2	**19**	**4**
A7	1	1	1	1	**28**	**1**	1	1	3	6	**21**	**3**	1	1	1	1	**28**	**1**
Best Alt.	**A7**						**A4**						**A7**					

Note: The values of UDS and UDR (under UDM) and achieved alternatives are shown in bold text.

The four considered MCDM techniques have evaluated across each alternative in TPE and CPE separately. In the case of OPE, all criteria except substation capacity are utilized in further evaluation with MCDM techniques. It is worth mentioning that in the all calculations of CPE and OPE cases, values of AIC were considered as achieved in [31].

4.1.2. MCDM Evaluation of Alternatives in Case 1 Under Normal Load (C1_NL)

The MCDM evaluations for C1_NL are illustrated in Figure 5a–d. The detailed numerical evaluations can be found in the supplementary file. Here, only the preference scores were illustrated in the figures to sort out a best alternative on the basis of TPE, CPE, and OPE evaluation per MCDM technique against each respective scenario. Refer to Table 5 for respective TPE and CPE in terms of numerical details without normalization. The rank of the alternatives in C1_NL are shown in Table 6.

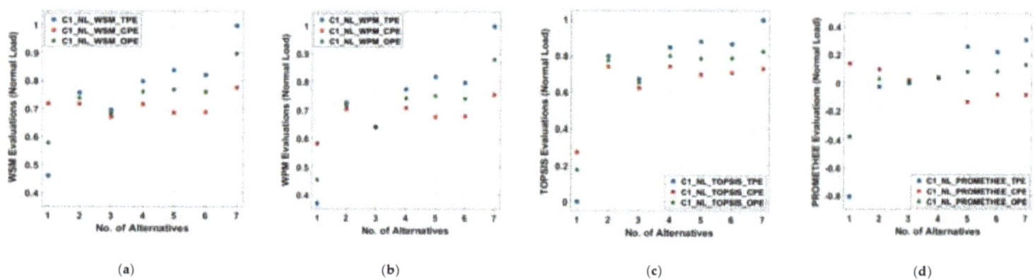

Figure 5. MCDM evaluations for C1_NL in 33-bus MDN: (**a**) weighted sum method (WSM) scores; (**b**) weighted product method (WPM) scores; (**c**) technique for order preference by similarity to ideal solution (TOPSIS) scores; (**d**) preference ranking organization method for enrichment of evaluations (PROMETHEE) scores.

As per the results, initially, the best alternative in TPE and OPE based evaluations is A7 as highlighted in Table 6, whereas in CPE there is no unanimous optimal solution. It is also observed that change of rank is more visible in CPE based evaluations with every MCDM approach.

After applying UDM with respective score and resulting alternative rank, the unanimous best solution utilizing various MCDM methodologies across TPE, CPE, and OPE are A7 (UDS=28 and UDR=1), A4 (UDS=22 and UDR=1), and A7 (UDS=28 and UDR=1); respectively. The UDM with UDS and UDR across TPE, CPE and OPE in C1_NL have highlighted in bold text as shown in Table 6.

4.2. Case 2 Under Normal Load (C2_NL): DGs Only Assets Placements in MDN Operating at 0.85 LPF

4.2.1. Initial Evaluation of Alternatives in Case 2 Under Normal Load (C2_NL)

The initial evaluation of case 2 (C2_NL) for each alternative is shown in terms of TPE and CPE is shown in Table 6. The numerical results were evaluated for seven alternatives using DG only asset placement operating at 0.85 LPF under NL.

The MCDM techniques were evaluated across each alternative in TPE and CPE in the same manner as in C1. The main reason for considering the 0.85 LPF of DG is close to that of DN is that it provides more reactive power support at load centers compared to those DGs operating at 0.90 LPF. It is advocated in various publications that such an arrangement results in achieving a better system LM with considerably better VP.

4.2.2. MCDM Evaluation of Alternatives in Case 2 under Normal Load (C2_NL)

The MCDM evaluations for case 2 (C2_NL) are illustrated in Figure 6a–d. The evaluation in terms of best and worst alternative with respective scores are shown below.

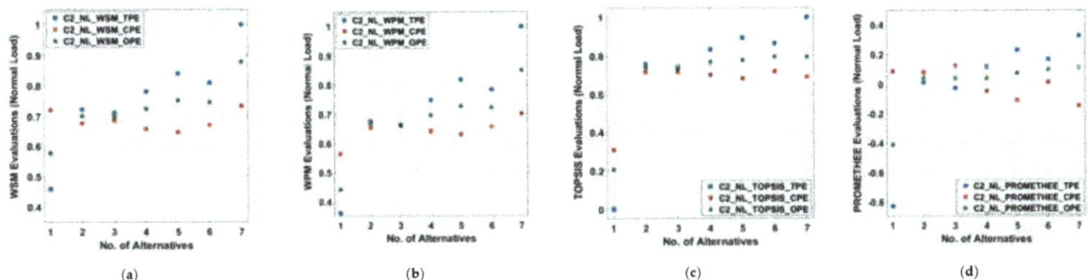

Figure 6. MCDM evaluations for C2_NL in 33-bus MDN: (**a**) WSM scores; (**b**) WPM scores; (**c**) TOPSIS scores; (**d**) PROMETHEE scores.

Refer to Table 7 for respective TPE and CPE in terms of numerical details without normalization. The rank of the alternatives in C2_NL is shown in Table 8. As per MCDM results, the best alternative in TPE is A7, whereas no unanimous solution is obtained in CPE and OPE. However, they are dominated by A7. After applying UDM with the best solution in TPE, CPE, and OPE are A7 (UDS=28 and UDR=1), A3 (UDS=23 and UDR=1), and A7 (UDS=27 and UDR=1); respectively. The UDM with UDS and UDR across TPE, CPE and OPE in C2_NL have highlighted in bold text as shown in Table 8.

Table 7. Techno-economic evaluation analysis in case 2 (C2_NL) for 33-bus MDN.

S#:		Technical Parameters Evaluations (TPE)							Cost (Economics Related) Parameters Evaluations (CPE)						
Case (No.) / Alt. (No).	DG Size (KVA) @ Bus Loc.	P_{Loss} (KW)	Q_{Loss} (KVAR)	*PLM* (%)	*QLM* (%)	*DGPP* (%)	V_{Min} (P.U)	$P_{SSR} + j\,Q_{SSR}$ (KW) + j(KVAR)	*PLC* (M$)	*PLS* (M$)	*CPDG* ($/MWh)	*CQDG* ($/MVArh)	*AIC* (M$)	*ACD* (M$)
C2/A1	DG1: 1970@15	115.6	79.7	45.21	44.27	45.06	0.9620	2505 + j 1604	0.0608	0.0501	34.471	4.5174	0.3558	0
C2/A2	DG1: 2709@30	35.2	28.98	83.32	79.73	61.993	0.9761	1447.7 + j 902	0.0185	0.0924	46.298	7.2777	0.4895	0
C2/A3	DG1: 950@15 DG2: 1633@30	38.3	28.1	81.85	80.35	59.12	0.9719	1557.8 + j 967.42	0.0201	0.0908	44.161	6.94	0.46673	0
C2/A4	DG1: 1886@30 DG2: 1167@25	27.93	24.39	86.76	82.94	69.874	0.9773	1147.7 + j 716.578	0.01468	0.09622	52.153	8.203	0.5517	0
C2/A5	DG1: 828.3@15 DG2: 1644@30 DG3: 727.8@7	26.7	16.75	87.35	88.29	73.24	0.9904	1021.6 + j 631	0.0140	0.0969	54.6517	8.598	0.57824	0
C2/A6	DG1: 877@15 DG2: 1310@30 DG3: 725@25	28.8	17.81	86.35	87.55	66.65	0.9876	1268.6 + j 783.81	0.0151	0.0958	49.754	7.8239	0.52618	0
C2/A7	DG1: 1422@30 DG2: 1045@25 DG3: 933.4@8	13.85	11.5	93.44	91.96	77.834	0.988	838.5 + j 520	0.00728	0.10362	58.0651	9.1375	0.6145	0

Table 8. Order of the ranks across TPE, CPE, and OPE in C2_NL for 33-bus MDN.

Evaluations	TPE (C2_NL)						CPE (C2_NL)						OPE (C2_NL)					
C2_NL / Alt (#)	WSM	WPM	TOPSIS	PROMETHEE	UDS	UDR	WSM	WPM	TOPSIS	PROMETHEE	UDS	UDR	WSM	WPM	TOPSIS	PROMETHEE	UDS	UDR
A1	7	7	7	7	**4**	**7**	2	7	7	2	**14**	**5**	7	7	7	7	**4**	**7**
A2	5	5	5	5	**12**	**5**	4	4	2	3	**19**	**3**	5	5	5	4	**13**	**5**
A3	6	6	6	6	**8**	**6**	3	2	3	1	**23**	**1**	6	6	6	5	**9**	**6**
A4	4	4	4	4	**16**	**4**	6	5	4	5	**12**	**6**	4	4	4	6	**14**	**4**
A5	2	2	2	2	**24**	**2**	7	6	6	6	**11**	**7**	2	2	3	3	**22**	**3**
A6	3	3	3	3	**20**	**3**	5	3	1	4	**19**	**2**	3	3	1	2	**23**	**2**
A7	1	1	1	1	**28**	**1**	1	1	5	7	**18**	**4**	1	1	2	1	**27**	**1**
Best Alt.	**A7**						**A3**						**A7**					

Note: The values of UDS and UDR (under UDM) and achieved alternatives are shown in bold text.

4.3. *Case 3 Under Normal Load (C3_NL): Asset Set (REG + D-STATCOM) Placements Equivalent to 0.90 LPF*

4.3.1. Initial Evaluation of Alternatives in Case 3 under Normal Load (C3_NL)

The initial evaluation of case 3 (C3_NL) for each alternative in terms of TPE and CPE is shown in Table 9. The numerical values refer to evaluated indices values for seven alternatives utilizing an asset set of REG (i.e., PV) and D-STATCOM for providing active and reactive power source, which is equal to single DG operating at 0.90 LPF under NL. The MCDM techniques were evaluated across each alternative is in the same manner as previous cases.

Table 9. Techno-economic evaluation analysis in case 3 (C3_NL) for 33-bus MDN.

S#:		Technical Parameters Evaluations (TPE)							Cost (Economics Related) Parameters Evaluations (CPE)					
Case (No.) / Alt. (No).	DG Size (KVA) @ Bus Loc.	P_{Loss} (KW)	Q_{Loss} (KVAR)	*PLM* (%)	*QLM* (%)	*DGPP* (%)	V_{Min} (P.U)	P_{SSR} + j Q_{SSR} (KW) + j(KVAR)	*PLC* (M $)	*PLS* (M $)	*CPDG* ($/MWh)	*CQDG* ($/MVArh)	*AIC* (M$)	*ACD* (M $)
C3/A1	S1: 1536 + j744 @ 15	85.59	50.03	58.65	64.32	39.067	0.9724	2264.6 + j 1606.3	0.0499	0.0659	30.99	3.0344	0.2901	0.00981
C3/A1	S1: 2475 + j1199 @ 30	39	29.38	81.52	79.45	62.94	0.9764	+ 1279 j 1130.4	0.0205	0.0904	49.75	4.9533	0.4675	0.01581
C3/A3	S1: 869.2 + j421.2 @ 15 S2: 1604 + j777.4 @ 30	27.89	16.20	86.52	88.44	62.89	0.9874	1269.7 + j 1117.6	0.0147	0.0962	49.71	4.9558	0.4672	0.01589
C3/A4	S1: 2121 + j1028 @ 30 S2: 486 + j236 @ 25	34	26.09	83.89	81.25	66.31	0.9768	1142 + j 1062.1	0.0179	0.093	52.39	5.227	0.4925	0.01666
C3/A5	S1: 620.5 + j300.5 @15 S2: 1442 + j698.3 @ 30 S3: 637.5 + j308.73 @ 7	19.40	11.09	90.63	92.09	68.65	0.99	1034 + j 1004.1	0.0102	0.1007	54.25	5.3593	0.5099	0.01718
C3/A6	S1: 789 + j380.7 @ 15 S2: 1247 + j586.2 @ 30 S3: 739.6 + j372 @ 25	20.11	12.23	90.29	91.28	70.53	0.9889	958.2 + j 973.33	0.0106	0.1003	55.76	5.507	0.5243	0.01766
C3/A7	S1: 1761 + j853 @ 30 S2: 450 + j218 @ 25 S3: 684 + j331.3 @ 8	20.8	13.76	90.14	90.37	73.63	0.9856	840.8 + j 911.46	0.0109	0.0999	58.16	5.7941	0.5467	0.01849

4.3.2. MCDM Evaluation of Alternatives in Case 3 Under Normal Load (C3_NL)

The MCDM evaluations for case 3 (C3_NL) are illustrated in Figure 7a–d. Refer to Table 9 for respective TPE and CPE in terms of numerical details without normalization. The rank of the alternatives in C3_NL is shown in Table 10.

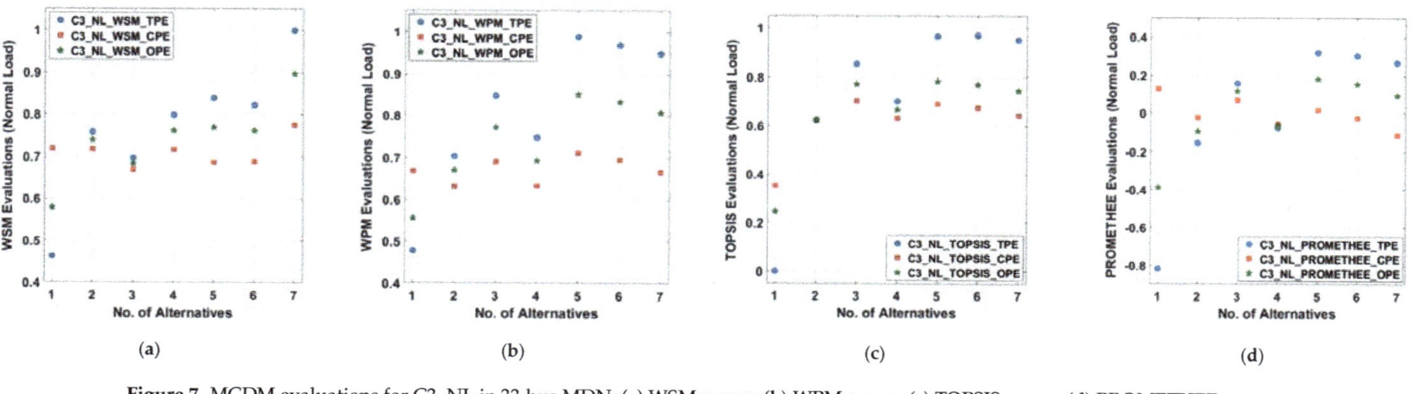

Figure 7. MCDM evaluations for C3_NL in 33-bus MDN: (**a**) WSM scores; (**b**) WPM scores; (**c**) TOPSIS scores; (**d**) PROMETHEE scores.

Table 10. Order of the ranks across TPE, CPE, and OPE in C3_NL for 33-bus MDN.

Evaluations	TPE (C3_NL)						CPE (C3_NL)						OPE (C3_NL)					
C3_NL / Alt (#)	WSM	WPM	TOPSIS	PROMETHEE	UDS	UDR	WSM	WPM	TOPSIS	PROMETHEE	UDS	UDR	WSM	WPM	TOPSIS	PROMETHEE	UDS	UDR
A1	7	7	7	7	**4**	**7**	1	1	7	1	**22**	**2**	7	7	7	7	**4**	**7**
A2	6	6	6	6	**8**	**6**	6	7	6	4	**9**	**7**	6	6	6	6	**8**	**6**
A3	4	4	4	4	**16**	**4**	4	2	1	2	**23**	**1**	4	4	2	3	**19**	**3**
A4	5	5	5	5	**12**	**5**	7	6	5	5	**9**	**6**	5	5	5	5	**12**	**5**
A5	1	1	1	1	**28**	**1**	2	4	3	7	**16**	**4**	1	1	3	1	**26**	**1**
A6	2	2	2	2	**24**	**2**	3	3	2	3	**21**	**3**	2	2	1	2	**25**	**2**
A7	3	3	3	3	**20**	**3**	5	5	4	6	**12**	**5**	3	3	4	4	**18**	**4**
Best Alt.				**A5**						**A3**						**A5**		

Note: The values of UDS and UDR (under UDM) and achieved alternatives are shown in bold text.

As per the results, the best alternative is A5 in OPE only. After applying UDM, the best solutions in TPE, CPE, and OPE were A5 (UDS=28 and UDR=1), A3 (UDS=23 and UDR=1), and A5 (UDS=26 and UDR=1); respectively. The UDM with UDS and UDR across TPE, CPE and OPE in C3_NL have highlighted in bold text as shown in Table 10.

4.4. Case 4 Under Normal Load (C4_NL): Asset Set (REG + D-STATCOM) Placements Equivalent to 0.85 LPF

4.4.1. Initial Evaluation of Alternatives in Case 4 under Normal Load (C4_NL)

The initial evaluation of case 4 (C4_NL) for each alternative is shown in terms of TPE and CPE is shown in Table 11. The numerical values were evaluated for seven alternatives utilizing an asset set of renewable DG i.e., PV and D-STATCOM for providing active and reactive power source, which is equal to single DG operating at 0.85 LPF under NL.

The MCDM techniques were evaluated across each alternative in the same manner as previous cases as illustrated in Figure 8a–d.

Table 11. Techno-economic evaluation analysis in case 4 (C4_NL) for 33-bus MDN.

S#:		Technical Parameters Evaluations (TPE)							Cost (Economics Related) Parameters Evaluations (CPE)					
Case (No.) / Alt. (No).	DG Size (KVA) @ Bus Loc.	P_{Loss} (KW)	Q_{Loss} (KVAR)	*PLM* (%)	*QLM* (%)	*DGPP* (%)	V_{Min} (P.U)	P_{SSR} + j Q_{SSR} (KW) + j (KVAR)	*PLC* (M $)	*PLS* (M $)	*CPDG* ($/MWh)	*CQDG* ($/MVArh)	*AIC* (M $)	*ACD* (M $)
C4/A1	S1: 1407.4 + j872 @ 15	84.41	56.24	59.22	59.89	37.89	0.9718	2392 + j 1484.24	0.0444	0.0665	28.39	4.4226	0.2658	0.01147
C4/A2	S1: 2302 + j1427 @ 30	36	29.7	82.94	79.23	61.97	0.9760	1449 + j 902.7	0.0189	0.0919	46.29	7.2694	0.4348	0.01881
C4/A3	S1: 807.5 + j485 @ 15 S2: 1388 + j893@ 30	27.39	17.98	86.77	87.18	59.32	0.9859	1546.9 + j 940	0.0144	0.0965	44.16	7.1574	0.4147	0.01821
C4/A4	S1: 1604 + j994 @ 30 S2: 992.5 + j615 @ 25	29.3	25.1	86.11	82.45	69.9	0.9772	1147.8 + j 709	0.0154	0.0955	52.16	8.2219	0.4905	0.02124
C4/A5	S1: 547 + j338.8 @15 S2: 1397 + j866 @ 30 S3: 606.3 + j376 @ 7	17.33	11.37	91.62	91.89	68.67	0.9901	1182 + j 730.57	0.0091	0.1018	51.25	8.0011	0.4817	0.02077
C4/A6	S1: 746 + j462 @ 15 S2: 1114 + j690@ 30 S3: 616 + j382 @ 25	20.38	13.51	90.16	90.37	66.64	0.9877	1259.4 + j 779.5	0.0107	0.1002	49.77	7.8765	0.4677	0.02029
C4/A7	S1: 1210 + j750 @ 30 S2: 890 + j551.2 @ 25 S3: 793.7 + j492 @ 8	16.3	12.6	92.27	91.19	77.89	0.9878	838.06 + j 519.4	0.0086	0.1023	58.07	9.1760	0.54653	0.02368

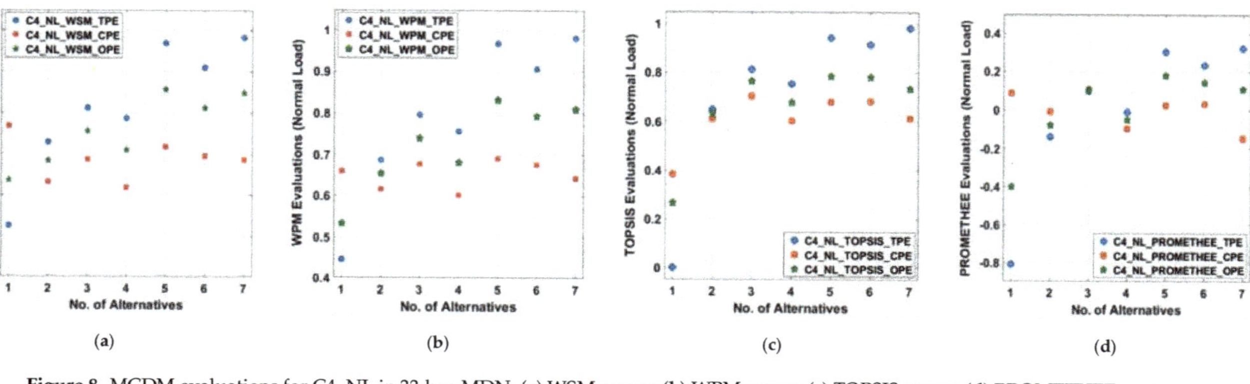

Figure 8. MCDM evaluations for C4_NL in 33-bus MDN: (**a**) WSM scores; (**b**) WPM scores; (c) TOPSIS scores; (**d**) PROMETHEE scores.

4.4.2. MCDM Evaluation of Alternatives in Case 4 Under Normal Load (C4_NL)

The MCDM evaluations for case 4 are illustrated in Figure 8a–d. Refer to Table 11 for respective TPE and CPE in terms of numerical details without normalization.

The rank of the alternatives in C4_NL is shown in Table 12. As per the results, the best alternative is A7 in TPE and A5 in OPE. After applying UDM, the best solutions in TPE, CPE, and OPE were A7 (UDS=28 and UDR=1), A3 (UDS=24 and UDR=1), and A5 (UDS=28 and UDR=1); respectively. The UDM with UDS and UDR across TPE, CPE and OPE in C4_NL have highlighted in bold text as shown in Table 12.

Table 12. Order of the ranks across TPE, CPE, and OPE in C4_NL for 33-bus MDN.

Evaluations	TPE (C4_NL)						CPE (C4_NL)						OPE (C4_NL)					
C4_NL / Alt (#)	WSM	WPM	TOPSIS	PROMETHEE	UDS	UDR	WSM	WPM	TOPSIS	PROMETHEE	UDS	UDR	WSM	WPM	TOPSIS	PROMETHEE	UDS	UDR
A1	7	7	7	7	**4**	**7**	1	4	7	2	**18**	**4**	7	7	7	7	**4**	**7**
A2	6	6	6	6	**8**	**6**	6	6	5	5	**10**	**6**	6	6	6	6	**8**	**6**
A3	4	4	4	4	**16**	**4**	4	2	1	1	**24**	**1**	4	4	3	4	**17**	**4**
A4	5	5	5	5	**12**	**5**	7	7	6	6	**6**	**7**	5	5	5	5	**12**	**5**
A5	2	2	2	2	**24**	**2**	2	1	3	4	**22**	**2**	1	1	1	1	**28**	**1**
A6	3	3	3	3	**20**	**3**	3	3	2	3	**21**	**3**	3	3	2	2	**22**	**2**
A7	1	1	1	1	**28**	**1**	5	5	4	7	**11**	**5**	2	2	4	3	**21**	**3**
Best Alt.	**A7**						**A3**						**A5**					

Note: The values of UDS and UDR (under UDM) and achieved alternatives are shown in bold text.

4.5. Case 1 Under Load Growth (C1_LG): DGs Only Assets Placements in MDN Operating at 0.90 LPF

4.5.1. Initial Evaluation of Alternatives in Case 1 Under Load Growth (C1_LG)

The initial evaluation from the perspective of TPE and CPE, of C1_LG, is shown in Table 13. The case of load growth refers to the condition, in which asset sitting and sizing achieved during normal load is kept constant across a planning horizon and load is incremented annually at a rate of 7.5%. The main aim of this evaluation is to find the change in the rank of alternatives initially evaluated as optimal ones.

The MCDM evaluation under LG shows a dip in achieved preference score, as well as a change in the ranks of the alternatives. This analysis indicates the change of rank and respective impacts on the active MDN across the LG. Also, the initially optimal solution may not remain feasible and the sub-optimal one may become a better choice, after a certain planning horizon.

Table 13. Techno-economic evaluation analysis in case 1 (C1_LG) for 33-bus MDN.

S#:		Technical Parameters Evaluations (TPE)							Cost (Economics Related) Parameters Evaluations (CPE)					
Case (No.) / Alt. (No).	DG Size (KVA) @ Bus Loc.	P_{Loss} (KW)	Q_{Loss} (KVAR)	*PLM* (%)	*QLM* (%)	*DGPP* (%)	V_{Min} (P.U)	P_{SSR} + j Q_{SSR} (KW) + j(KVAR)	*PLC* (M $)	*PLS* (M $)	*CPDG* ($/MWh)	*CQDG* ($/MVArh)	*AIC* (M $)	*ACD* (M $)
C1/A1	DG1: 2013@15	228.88	113.73	49.211	62.73	27.214	0.9475	4110.93 + j 2516.51	0.4648	0.5398	30.97	3.0773	0.3084	0
C1/A2	DG1: 2750@30	127.95	75.356	71.61	75.31	43.842	0.9481	2985.95 + j 2178.7	0.2588	0.7458	49.75	4.9527	0.4969	0
C1/A3	DG1: 971@15 DG2: 1783@30	95.35	53.664	78.842	82.42	43.802	0.9641	2949.75 + j 2155.224	0.1762	0.8284	48.80	4.8567	0.4965	0
C1/A4	DG1: 2357@30 DG2: 540@25	117.93	75.356	73.83	77.09	46.185	0.9485	2843.63 + j 2114.583	0.2328	0.7718	52.39	5.2141	0.5235	0
C1/A5	DG1: 832.6@15 DG2: 1602@30 DG3: 745.1@7	83.661	46.815	81.435	84.66	47.822	0.9638	2554.93 + j 1962.8	0.1399	0.8647	54.24	5.4023	0.5420	0
C1/A6	DG1: 894.6@15 DG2: 1386@30 DG3: 822.6@25	77.621	44.204	82.776	85.51	49.131	0.9652	2617.7 + j 1993.55	0.1361	0.8684	55.77	5.4988	0.5569	0
C1/A7	DG1: 1957@30 DG2: 500@25 DG3: 760@8	81.611	47.31	81.89	84.49	51.287	0.9576	2519.3 + j 1946.7	0.1458	0.8587	58.16	5.7938	0.5813	0

4.5.2. MCDM Evaluation of Alternatives in Case 1 under Load Growth (C1_LG)

The MCDM evaluations for case 1 are illustrated in Figure 9a–d. The rank of the alternatives in C1_LG is shown in Table 14. As per the results, the best alternative in TPE is A6, whereas there is not a unanimous solution in CPE and OPE, respectively. After applying UDM, the best solutions in TPE, CPE, and OPE are A6 (UDS=28 and UDR=1), A3 (UDS=23 and UDR=1), and A6 (UDS=26 and UDR=1); respectively. The UDM with UDS and UDR across TPE, CPE and OPE in C1_LG have highlighted in bold text as shown in Table 14.

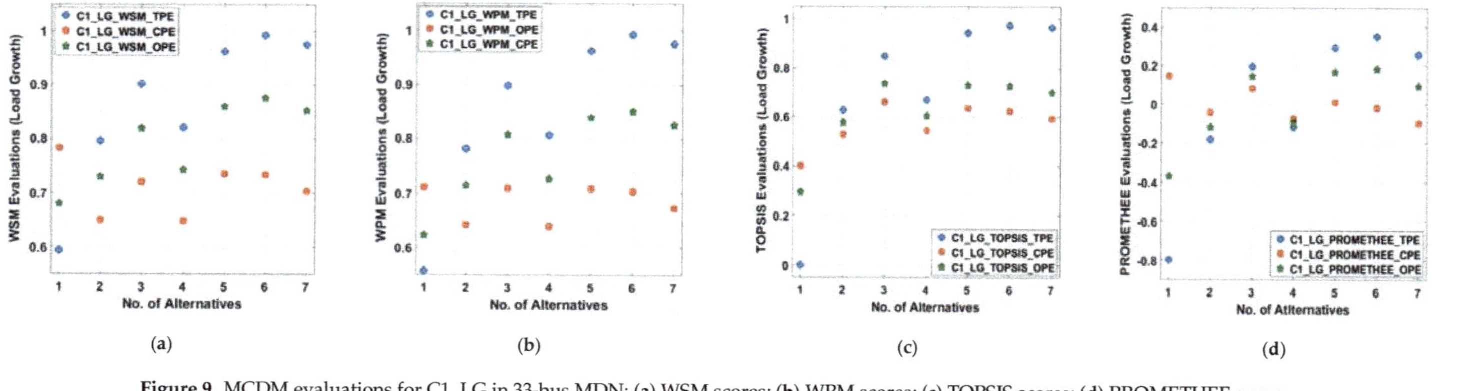

Figure 9. MCDM evaluations for C1_LG in 33-bus MDN: (**a**) WSM scores; (**b**) WPM scores; (**c**) TOPSIS scores; (**d**) PROMETHEE scores.

Table 14. Order of the ranks across TPE, CPE, and OPE in C1_LG for 33-bus MDN.

Evaluations	TPE (C1_LG)						CPE (C1_LG)						OPE (C1_LG)					
C1_LG / Alt (#)	WSM	WPM	TOPSIS	PROMETHEE	UDS	UDR	WSM	WPM	TOPSIS	PROMETHEE	UDS	UDR	WSM	WPM	TOPSIS	PROMETHEE	UDS	UDR
A1	7	7	7	7	**4**	**7**	1	1	7	1	**22**	**2**	7	7	7	7	**4**	**7**
A2	6	6	6	6	**8**	**6**	6	7	6	5	**8**	**7**	6	6	6	6	**8**	**6**
A3	4	4	4	4	**16**	**4**	4	2	1	2	**23**	**1**	4	4	1	3	**20**	**3**
A4	5	5	5	5	**12**	**5**	7	6	5	6	**8**	**6**	5	5	5	5	**12**	**5**
A5	3	3	3	2	**21**	**3**	2	3	2	3	**22**	**3**	2	2	2	2	**24**	**2**
A6	1	1	1	1	**28**	**1**	3	4	3	4	**18**	**4**	1	1	3	1	**26**	**1**
A7	2	2	2	3	**23**	**2**	5	5	4	7	**11**	**5**	3	3	4	4	**18**	**4**
Best Alt.	**A6**						**A3**						**A6**					

Note: The values of UDS and UDR (under UDM) and achieved alternatives are shown in bold text.

4.6. Case 2 under Load Growth (C2_LG): DGs Only Assets Placements in MDN Operating at 0.85 LPF

4.6.1. Initial Evaluation of Alternatives in Case 2 under Load Growth (C2_LG)

The initial evaluation of case 2 (C2_LG) for each alternative under load growth in terms of TPE and CPE is shown in Table 15.

Table 15. Techno-economic evaluation analysis in case 2 (C2_LG) for 33-bus MDN.

S#:		Technical Parameters Evaluations (TPE)							Cost (Economics Related) Parameters Evaluations (CPE)					
Case (No.) / Alt. (No).	DG Size (KVA) @ Bus Loc.	P_{Loss} (KW)	Q_{Loss} (KVAR)	*PLM* (%)	*QLM* (%)	*DGPP* (%)	V_{Min} (P.U)	P_{SSR} + j Q_{SSR} (KW) + j(KVAR)	*PLC* (M$)	*PLS* (M$)	*CPDG* ($/MWh)	*CQDG* ($/MVArh)	*AIC* (M$)	*ACD* (M$)
C2/A1	DG1: 1970@15	204.55	131.47	54.61	56.92	26.39	0.9469	3863 + j2395.71	0.4352	0.5694	28.39	4.4488	0.2992	0
C2/A2	DG1: 2709@30	127.56	85.03	71.69	72.14	43.18	0.9478	3157.9 + j 1960	0.2559	0.7487	46.29	7.2777	0.4895	0
C2/A3	DG1: 950@15 DG2: 1633@30	101.56	65.00	77.44	78.69	41.18	0.9624	3239 + j 2006.32	0.1792	0.8254	44.16	7.1279	0.4667	0
C2/A4	DG1: 1886@30 DG2: 1167@25	109.27	73.91	75.75	75.78	48.67	0.949	2847.2 + j 1767.24	0.2139	0.7907	52.15	8.203	0.5517	0
C2/A5	DG1: 828.3@15 DG2: 1644@30 DG3: 727.8@7	81.44	51.93	81.93	82.98	47.83	0.9641	2694.36 + j 1669	0.1349	0.8697	51.26	8.0612	0.5421	0
C2/A6	DG1: 877@15 DG2: 1310@30 DG3: 725@25	84.72	54.55	81.2	82.12	46.42	0.9635	2942.5 + j 1822.55	0.1444	0.8602	49.75	7.4566	0.5262	0
C2/A7	DG1: 1422@30 DG2: 1045@25 DG3: 933.4@8	71.17	46.64	84.21	84.72	54.22	0.9597	2513.8 + j 1557.34	0.1243	0.8802	58.07	9.1375	0.6145	0

4.6.2. MCDM Evaluation of Alternatives in Case 2 under Load Growth (C2_LG)

The MCDM evaluations for case 2 (C2) under load growth are illustrated in Figure 10a–d. The rank of the alternatives in C2 under LG (C2_LG) is shown in Table 16. As per the results, the best alternative in TPE is A7, whereas no unanimous solution is obtained in CPE and OPE. After applying UDM, the best solutions in TPE, CPE, and OPE are A7 (UDS=28 and UDR=1), A3 (UDS=24 and UDR=1), and A5 (UDS=25 and UDR=1); respectively. The UDM with UDS and UDR across TPE, CPE and OPE in C2_LG have highlighted in bold text as shown in Table 16.

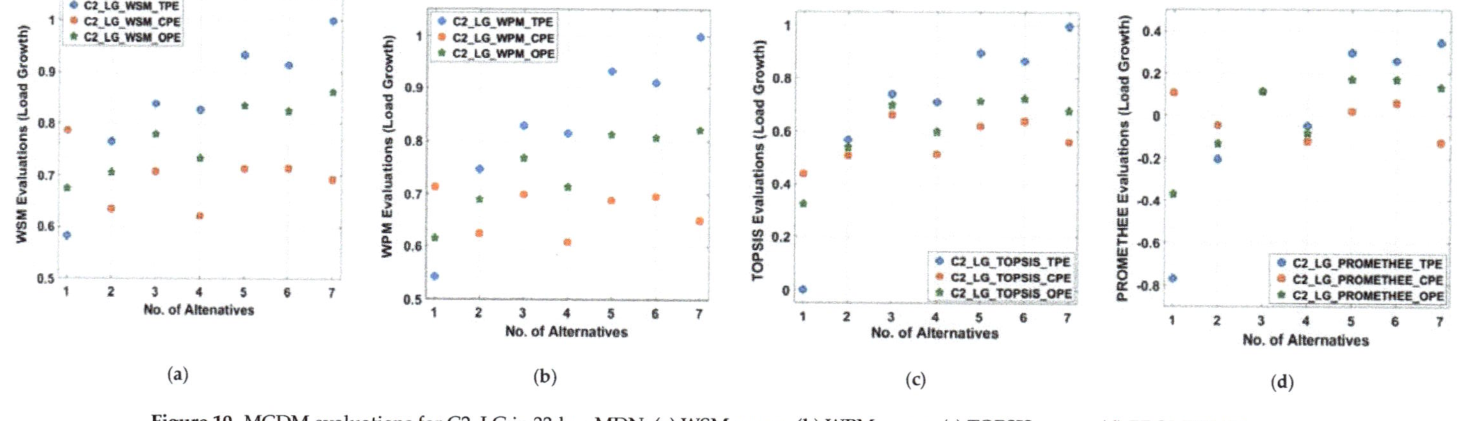

Figure 10. MCDM evaluations for C2_LG in 33-bus MDN: (**a**) WSM scores; (**b**) WPM scores; (**c**) TOPSIS scores; (**d**) PROMETHEE scores.

Table 16. Order of the ranks across TPE, CPE, and OPE in C2_LG for 33-bus MDN.

Evaluations	TPE (C2_LG)						CPE (C2_LG)						OPE (C2_LG)					
C2_LG / Alt (#)	WSM	WPM	TOPSIS	PROMETHEE	UDS	UDR	WSM	WPM	TOPSIS	PROMETHEE	UDS	UDR	WSM	WPM	TOPSIS	PROMETHEE	UDS	UDR
A1	7	7	7	7	**4**	**7**	1	1	7	2	**21**	**3**	7	7	7	7	**4**	**7**
A2	6	6	6	6	**8**	**6**	6	6	6	5	**9**	**6**	6	6	6	6	**8**	**6**
A3	4	4	4	4	**16**	**4**	4	2	1	1	**24**	**1**	4	4	3	4	**17**	**4**
A4	5	5	5	5	**12**	**5**	7	7	5	6	**7**	**7**	5	5	5	5	**12**	**5**
A5	2	2	2	2	**24**	**2**	3	4	3	4	**18**	**4**	2	2	2	1	**25**	**1**
A6	3	3	3	3	**20**	**3**	2	3	2	3	**22**	**2**	3	3	1	2	**23**	**3**
A7	1	1	1	1	**28**	**1**	5	5	4	7	**11**	**5**	1	1	4	3	**23**	**2**
Best Alt.	**A7**						**A3**						**A5**					

Note: The values of UDS and UDR (under UDM) and achieved alternatives are shown in bold text.

4.7. Case 3 Under Load Growth: Asset Set (REG + D-STATCOM) Placements Equivalent to 0.90 LPF (C3_LG)

4.7.1. Initial Evaluation of Alternatives in Case 3 under Normal Load (C3_LG)

The initial evaluation of C3_LG for each alternative in terms of TPE and CPE is shown in Table 11.

4.7.2. MCDM Evaluation of Alternatives in Case 3 under Load Growth (C3_LG)

The MCDM evaluations for C3_LG are illustrated in Figure 11a–d. Refer to Table 17 for respective TPE and CPE in terms of numerical details without normalization. The rank of the alternatives in C3 under LG is shown in Table 18. As per the results, the best alternative in TPE is A6, whereas there are no unanimous solutions in CPE and OPE, respectively. After applying UDM, the best solutions in TPE, CPE, and OPE are A6 (UDS=28 and UDR=1), A3 (UDS=23 and UDR=1), and A6 (UDS=26 and UDR=1), respectively. The UDM with UDS and UDR across TPE, CPE and OPE in C3_LG have highlighted in bold text as shown in Table 18.

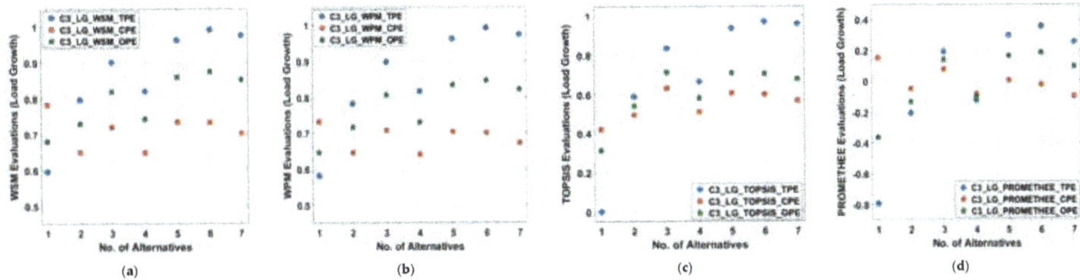

Figure 11. MCDM evaluations for C3_LG in 33-bus MDN: (**a**) WSM scores; (**b**) WPM scores; (**c**) TOPSIS scores; (**d**) PROMETHEE scores.

Table 17. Techno-economic evaluation analysis in case 3 (C3_LG) for 33-bus MDN.

S#:		Technical Parameters Evaluations (TPE)							Cost (Economics Related) Parameters Evaluations (CPE)					
Case (No.) / Alt. (No).	**DG Size (KVA) @ Bus Loc.**	P_{Loss} (KW)	Q_{Loss} (KVAR)	*PLM* (%)	*QLM* (%)	*DGPP* (%)	V_{Min} (P.U)	$P_{SSR} + j\, Q_{SSR}$ (KW) + j(KVAR)	*PLC* (M$)	*PLS* (M$)	*CPDG* ($/MWh)	*CQDG* ($/MVArh)	*AIC* (M$)	*ACD* (M$)
C3/A1	S1: 1536 + j744 @ 15	203.15	114.61	54.92	62.44	27.21	0.9474	4000 + j 2672.61	0.4342	0.5704	30.97	3.0735	0.2901	0.009812
C3/A1	S1: 2475 + j1199 @ 30	129.18	75.87	71.33	75.14	43.83	0.9481	2987.2 + j2178.9	0.2609	0.7436	49.25	4.9381	0.4675	0.015803
C3/A3	S1: 869.2 + j421.2 @ 15 S2: 1604 + j777.4 @ 30	96.74	54.28	78.53	82.21	43.81	0.964	2956.5 + j 2157.7	0.1789	0.8256	49.71	4.948	0.4672	0.01579
C3/A4	S1: 2121 + j1028 @ 30 S2: 486 + j236 @ 25	119.55	70.79	73.47	76.8	46.18	0.9484	2845.6 + j2108.8	0.2361	0.7684	52.39	5.2147	0.4924	0.016657
C3/A	S1: 620.5 + j300.5 @15 S2: 1442 + j698.3 @ 30 S3: 637.5 + j308.73 @ 7	85.29	47.38	81.07	84.47	47.92	0.9637	2718.3 + j 2041.85	0.1453	0.8593	54.25	5.5187	0.5100	0.01742
C3/A6	S1: 789 + j380.7 @ 15 S2: 1247 + j586.2 @ 30 S3: 739.6 + j372 @ 25	78.98	44.87	82.47	85.29	49.21	0.9651	2636.38 + j 2008	0.1398	0.8648	55.76	5.6037	0.5243	0.01781
C3/A7	S1: 1761 + j853 @ 30 S2: 450 + j218 @ 25 S3: 684 + j331.3 @ 8	83.27	48.32	81.52	84.17	51.31	0.9574	2521.27 + j 1948	0.1503	0.8542	58.16	5.8296	0.5468	0.018482

Table 18. Order of the ranks across TPE, CPE, and OPE in C3_LG for 33-bus MDN.

Evaluations	TPE (C3_LG)						CPE (C3_LG)						OPE (C3_LG)					
C3_LG / Alt (#)	**WSM**	**WPM**	**TOPSIS**	**PROMETHEE**	**UDS**	**UDR**	**WSM**	**WPM**	**TOPSIS**	**PROMETHEE**	**UDS**	**UDR**	**WSM**	**WPM**	**TOPSIS**	**PROMETHEE**	**UDS**	**UDR**
A1	7	7	7	7	**4**	**7**	1	1	7	1	**22**	**2**	7	7	7	7	**4**	**7**
A2	6	6	6	6	**8**	**6**	6	6	6	5	**9**	**6**	6	6	6	6	**8**	**6**
A3	4	4	4	4	**16**	**4**	4	2	1	2	**23**	**1**	4	4	1	3	**20**	**3**
A4	5	5	5	5	**12**	**5**	7	7	5	6	**7**	**7**	5	5	5	5	**12**	**5**
A5	3	3	3	2	**21**	**3**	3	3	2	3	**21**	**3**	2	2	2	2	**24**	**2**
A6	1	1	1	1	**28**	**1**	2	4	3	4	**19**	**4**	1	1	3	1	**26**	**1**
A7	2	2	2	3	**23**	**2**	5	5	4	7	**11**	**5**	3	3	4	4	**18**	**4**
Best Alt.				**A6**						**A3**						**A6**		

Note: The values of UDS and UDR (under UDM) and achieved alternatives are shown in bold text.

4.8. Case 4 Under Load Growth: Assets (DG + D-STATCOM) Placements Equivalent to 0.85 LPF (C4_LG)

4.8.1. Initial Evaluation of Alternatives in Case 4 Under Load Growth (C4_LG)

The initial evaluation of case 4 (C4_LG) for each alternative under LG in terms of TPE and CPE is shown in Table 19.

Table 19. Techno-economic evaluation analysis in case 4 (C4_LG) for 33-bus MDN.

S#:		Technical Parameters Evaluations (TPE)							Cost (Economics Related) Parameters Evaluations (CPE)					
Case (No.) / Alt. (No).	**DG Size (KVA) @ Bus Loc.**	P_{Loss} (KW)	Q_{Loss} (KVAR)	*PLM* (%)	*QLM* (%)	*DGPP* (%)	V_{Min} (P.U)	P_{SSR} + j Q_{SSR} (KW) + j(KVAR)	*PLC* (M$)	*PLS* (M$)	*CPDG* ($/MWh)	*CQDG* ($/MVArh)	*AIC* (M$)	*ACD* (M$)
C4/A1	S1: 1407.4 + j872 @ 15	203.48	130.92	54.85	57.09	26.36	0.9467	4129 + j 2560.92	0.4322	0.5724	28.39	4.4182	0.2658	0.011467
C4/A2	S1: 2302 + j1427 @ 30	127.85	85.07	71.63	72.13	43.18	0.9478	3158.9 + j 1960.07	0.2569	0.7477	46.29	7.2803	0.4348	0.018804
C4/A3	S1: 807.5 + j485 @ 15 S2: 1388 + j893@ 30	102.66	65.38	77.22	78.58	41.39	0.9623	3240.2 + j1989.4	0.1848	0.8198	44.16	7.2153	0.4147	0.018276
C4/A4	S1: 1604 + j994 @ 30 S2: 992.5 + j615 @ 25	110.79	74.57	75.42	75.57	48.71	0.9489	2847.3 + j 1767.57	0.2173	0.7873	52.16	8.2390	0.4904	0.021198
C4/A5	S1: 547 + j338.8 @15 S2: 1397 + j866 @ 30 S3: 606.3 + j376 @ 7	82.85	53.07	81.69	82.61	47.82	0.9638	2865.55 + j1774.27	0.1373	0.8673	51.25	8.0446	0.4817	0.020826
C4/A6	S1: 746 + j462 @ 15 S2: 1114 + j690@ 30 S3: 616 + j382 @ 25	85.79	54.16	80.96	82.25	46.42	0.9636	2942.8 + j1822.16	0.1483	0.8563	49.77	10.296	0.4677	0.020218
C4/A7	S1: 1210 + j750 @ 30 S2: 890 + j551.2 @ 25 S3: 793.7 + j492 @ 8	72.69	47.27	83.87	84.51	54.22	0.9595	2512 + j 1556.07	0.1289	0.8757	58.07	9.145	0.5466	0.023616

4.8.2. MCDM evaluation of alternatives in Case 4 under load growth (C4_LG)

The MCDM evaluations for C4_LG are illustrated in Figure 12a–d. The rank of the alternatives in C4_LG is shown in Table 20.

As per the results in Figure 12a–d and Table 20, the best alternative in TPE is A7, whereas there are no unanimous solutions in CPE and in OPE. After applying UDM, the best solutions in TPE, CPE, and OPE are A7 (UDS=28 and UDR=1), A3 (UDS=26 and UDR=1), and A5 (UDS=26 and UDR=1), respectively. The UDM with UDS and UDR across TPE, CPE and OPE in C4_LG have highlighted in bold text as shown in Table 20.

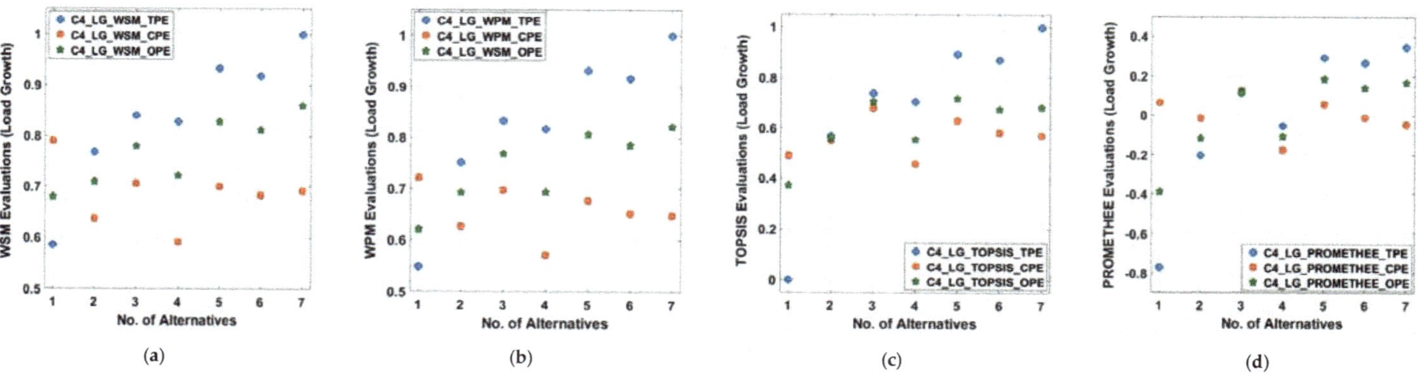

(a) (b) (c) (d)

Figure 12. MCDM evaluations for C4_LG in 33-bus MDN: (**a**) WSM scores; (**b**) WPM scores; (**c**) TOPSIS scores; (**d**) PROMETHEE scores.

Table 20. Order of the ranks across TPE, CPE, and OPE in C4_LG for 33-bus MDN.

Evaluations	TPE (C4_LG)						CPE (C4_LG)						OPE (C4_LG)					
C4_LG / Alt (#)	WSM	WPM	TOPSIS	PROMETHEE	UDS	UDR	WSM	WPM	TOPSIS	PROMETHEE	UDS	UDR	WSM	WPM	TOPSIS	PROMETHEE	UDS	UDR
A1	7	7	7	7	**4**	**7**	1	1	6	2	**22**	**2**	7	7	7	7	**4**	**7**
A2	6	6	6	6	**8**	**6**	6	6	5	5	**10**	**6**	6	6	6	6	**8**	**6**
A3	4	4	4	4	**16**	**4**	2	2	1	1	**26**	**1**	4	4	2	4	**18**	**4**
A4	5	5	5	5	**12**	**5**	7	7	7	7	**4**	**7**	5	5	5	5	**12**	**5**
A5	2	2	2	2	**24**	**2**	3	3	2	3	**21**	**3**	2	2	1	1	**26**	**1**
A6	3	3	3	3	**20**	**3**	5	4	3	4	**16**	**4**	3	3	4	3	**19**	**3**
A7	1	1	1	1	**28**	**1**	4	5	4	6	**13**	**5**	1	1	3	2	**25**	**2**
Best Alt.	**A7**						**A3**						**A5**					

Note: The values of UDS and UDR (under UDM) and achieved alternatives are shown in bold text.

4.9. *Case 1 Under Optimal Load Growth: DGs Only Assets Placements Operating at 0.90 LPF (C1_OLG)*

4.9.1. Initial Evaluation of Alternatives in Case 1 Under Optimal Load Growth (C1_OLG)

The initial evaluation of C1_OLG is shown in Table 21, from the perspective of TPE and CPE.

Table 21. Techno-economic evaluation analysis in case 1 (C1_OLG) for 33-bus MDN.

S#:		Technical Parameters Evaluations (TPE)							Cost (Economics Related) Parameters Evaluations (CPE)					
Case (No.) / Alt. (No).	**DG Size (KVA) @ Bus Loc.**	P_{Loss} (KW)	Q_{Loss} (KVAR)	*PLM* (%)	*QLM* (%)	*DGPP* (%)	V_{Min} (P.U)	$P_{SSR} + j\,Q_{SSR}$ (KW) + j(KVAR)	*PLC* (M$)	*PLS* (M$)	*CPDG* ($/MWh)	*CQDG* ($/MVArh)	*AIC* (M$)	*ACD* (M$)
C1/A1	DG1: 2205@15	178.15	111.24	60.47	63.55	39.06	0.9598	3526.62 + j2452.103	0.0936	0.9109	44.35	4.4124	0.4427	0
C1/A2	DG1: 3950@30	98.14	58.56	78.22	80.81	62.97	0.9656	1876.14 + j1638.795	0.0516	0.9530	71.35	7.1142	0.7137	0
C1/A3	DG1: 1500@15 DG2: 2300@30	60.34	35.29	86.62	88.43	60.58	0.9811	1973.34 + j1680.91	0.0317	0.9729	68.65	6.8440	0.6867	0
C1/A4	DG1: 3500@30 DG2: 590@25	87.3	54.63	80.63	82.09	65.21	0.9658	1739.3 + j1574.63	0.0459	0.9587	73.87	7.3659	0.7390	0
C1/A5	DG1: 980@15 DG2: 2235@30 DG3: 1521@7	34.99	21.80	92.23	92.86	75.52	0.9899	1105.59 + j1259.425	0.0184	0.9862	85.51	8.5309	0.8559	0
C1/A6	DG1: 1147@15 DG2: 2119@30 DG3: 1272@25	39.45	24.17	91.25	92.08	72.36	0.9845	1288.25 + j1347.487	0.0207	0.9839	81.95	8.1748	0.82	0
C1/A7	DG1: 2890@30 DG2: 590@25 DG3: 1090@8	45.68	28.21	89.86	90.75	72.86	0.9792	1265.65 + j1338.2	0.0240	0.9806	82.51	8.2303	0.8258	0

4.9.2. MCDM Evaluation of Alternatives in Case 1 Under Optimal Load Growth (C1_OLG)

The MCDM evaluations for case 1 are illustrated in Figure 13a–d. The rank of the alternatives in C1_OLG is shown in Table 22. As per the results, the best alternative in OPE is A5, whereas there is not a unanimous solution in TPE and CPE, respectively. After applying UDM, the best solutions in TPE, CPE, and OPE are A5 (UDS=24 and UDR=1), A3 (UDS=23 and UDR=1), and A5 (UDS=26 and UDR=1); respectively. The UDM with UDS and UDR across TPE, CPE and OPE in C1_OLG have highlighted in bold text as shown in Table 22.

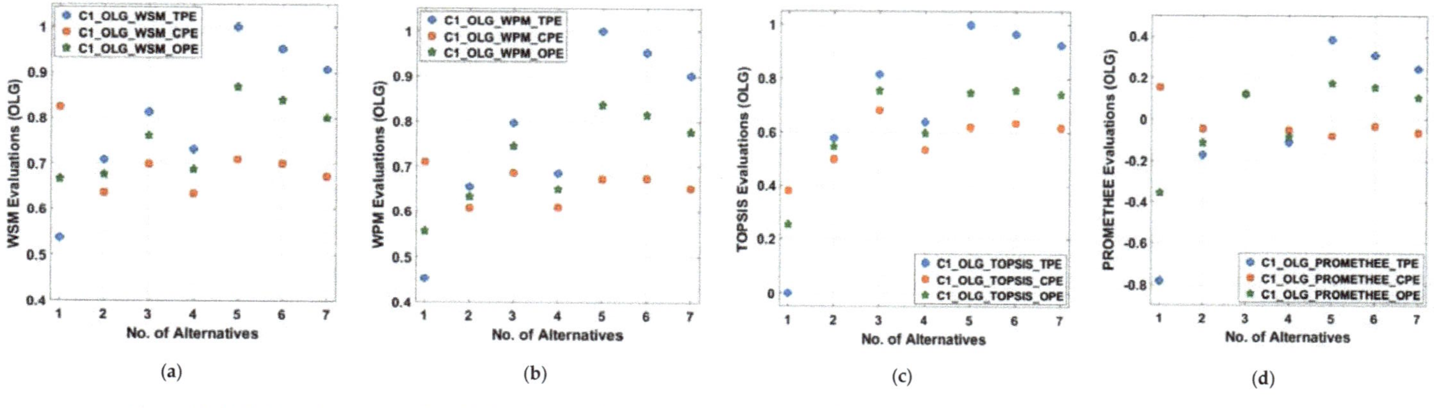

Figure 13. MCDM evaluations for C1_OLG in 33-bus MDN: (**a**) WSM scores; (**b**) WPM scores; (**c**) TOPSIS scores; (**d**) PROMETHEE scores.

Table 22. Order of the ranks across TPE, CPE, and OPE in C1_OLG for 33-bus MDN.

Evaluations	TPE (C1_OLG)						CPE (C1_OLG)						OPE (C1_OLG)					
C1_OLG / Alt (#)	WSM	WPM	TOPSIS	PROMETHEE	UDS	UDR	WSM	WPM	TOPSIS	PROMETHEE	UDS	UDR	WSM	WPM	TOPSIS	PROMETHEE	UDS	UDR
A1	7	7	7	7	**4**	**7**	1	1	7	1	**22**	**2**	7	7	7	7	**4**	**7**
A2	6	6	6	6	**8**	**6**	6	7	6	4	**9**	**7**	6	6	6	6	**8**	**6**
A3	4	4	4	4	**16**	**4**	4	2	1	2	**23**	**1**	4	4	2	3	**19**	**3**
A4	5	5	5	5	**12**	**5**	7	6	5	5	**9**	**6**	5	5	5	5	**12**	**5**
A5	1	1	1	1	**28**	**1**	2	4	2	7	**17**	**4**	1	1	3	1	**26**	**1**
A6	2	2	2	2	**24**	**2**	3	3	3	3	**20**	**3**	2	2	1	2	**25**	**2**
A7	3	3	3	3	**20**	**3**	5	5	4	6	**12**	**5**	3	3	4	4	**18**	**4**
Best Alt.				A5						A3						A5		

Note: The values of UDS and UDR (under UDM) and achieved alternatives are shown in bold text.

4.10. Case 2 under Optimal Load Growth: Dgs Only Assets Placements Operating at 0.85 LPF (C2_OLG)

4.10.1. Initial Evaluation of Alternatives in Case 2 under Optimal Load Growth (C2_OLG)

The initial evaluation of case 2 (C2_OLG) under OLG for each alternative in terms of TPE and CPE is shown in Table 23.

Table 23. Techno-economic evaluation analysis in case 2 (C2_OLG) for 33-bus MDN.

S#:		Technical Parameters Evaluations (TPE)							Cost (Economics Related) Parameters Evaluations (CPE)					
Case (No.) / Alt. (No).	**DG Size (KVA) @ Bus Loc.**	P_{Loss} (KW)	Q_{Loss} (KVAR)	*PLM* (%)	*QLM* (%)	*DGPP* (%)	V_{Min} (P.U)	P_{SSR} + j Q_{SSR} (KW) + j(KVAR)	*PLC* (M$)	*PLS* (M$)	*CPDG* ($/MWh)	*CQDG* ($/MVArh)	*AIC* (M$)	*ACD* (M$)
C2/A1	DG1: 2410@15	179.26	119.65	60.22	60.79	38.42	0.9594	3463.8 + j2152.104	0.09421	0.9104	41.22	6.4752	0.4355	0
C2/A2	DG1: 3925@30	95.79	66.52	78.74	78.2	62.52	0.9656	2092.54 + j1300.9	0.0504	0.9542	66.98	10.545	0.7092	0
C2/A3	DG1: 1390@15 DG2: 2450@30	55.75	36.81	87.63	87.93	61.22	0.9814	2124.75 + j1158.91	0.0293	0.9753	65.53	10.317	0.6939	0
C2/A4	DG1: 2825@30 DG2: 1500@25	76.79	54.97	82.96	81.99	68.95	0.9667	1733.54 + j1078.635	0.0404	0.9642	73.78	11.621	0.7815	0
C2/A5	DG1: 980@15 DG2: 2235@30 DG3: 1177@7	34.63	23.32	92.31	92.36	70.04	0.9878	1634.43 + j1011.7	0.0182	0.9864	74.94	11.804	0.7938	0
C2/A6	DG1: 1147@15 DG2: 2077@30 DG3: 1277@25	36.89	25.38	91.81	91.68	71.69	0.9847	1544.04 + j956.33	0.0194	0.9852	76.78	11.996	0.8125	0
C2/A7	DG1: 2100@30 DG2: 1400@25 DG3: 1360@8	36.42	25.32	91.92	91.7	77.48	0.9827	1238.42 + j767.156	0.0191	0.9854	82.87	13.057	0.8782	0

4.10.2. MCDM Evaluation of Alternatives in Case 2 under Optimal Load Growth (C2_OLG)

The MCDM evaluations for case 2 (C2_OLG) under optimal load growth are illustrated in Figure 14a–d. The rank of the alternatives in C2_OLG is shown in Table 24. As per the results, the best alternative in OPE is A5, whereas no unanimous solution is obtained in CPE and OPE. After applying UDM, the best solutions in TPE, CPE, and OPE are A5 (UDS=26 and UDR=1), A5 (UDS=23 and UDR=1), and A5 (UDS=28 and UDR=1), respectively.

It can be seen in Table 24 from CPE (C2_OLG) that the UDS score of A1 and A3 is the same (UDS=22). As per the aforementioned rules devised for UDM in Section 2.5, the solution with the highest number of highest priority ranks will be given preference. Hence, alternative A1 (UDR=2) is given preference over A3 (UDR=3) for the second-best alternative despite having the same score from the viewpoint of UDS. The UDM with UDS and UDR across TPE, CPE and OPE in C2_OLG have highlighted in bold text as shown in Table 24.

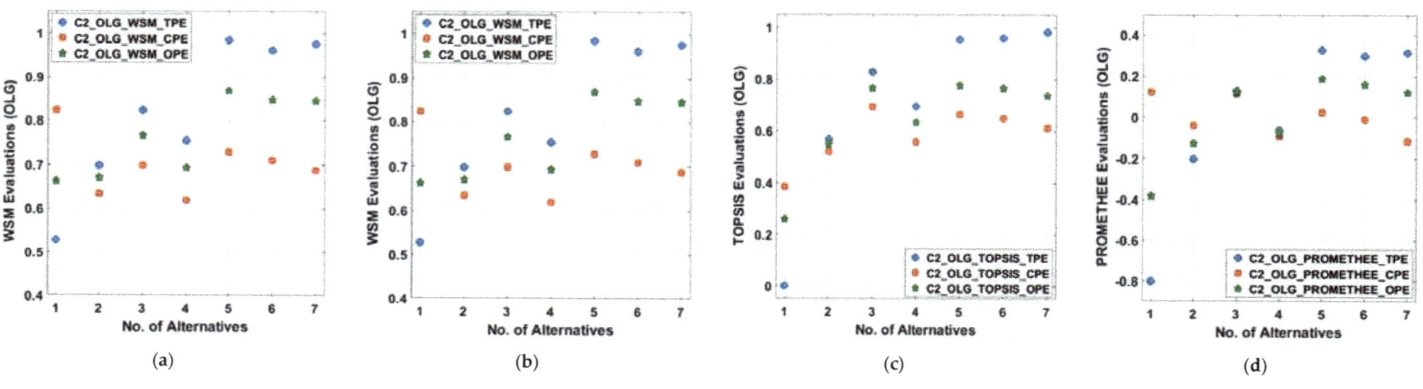

Figure 14. MCDM evaluations for C2_OLG in 33-bus MDN: (**a**) WSM scores; (**b**) WPM scores; (**c**) TOPSIS scores; (**d**) PROMETHEE scores.

Table 24. Order of the ranks across TPE, CPE, and OPE in C2_OLG for 33-bus MDN.

Evaluations	TPE (C2_OLG)						CPE (C2_OLG)						OPE (C2_OLG)					
C2_OLG / Alt (#)	WSM	WPM	TOPSIS	PROMETHEE	UDS	UDR	WSM	WPM	TOPSIS	PROMETHEE	UDS	UDR	WSM	WPM	TOPSIS	PROMETHEE	UDS	UDR
A1	7	7	7	7	**4**	**7**	1	1	7	1	**22**	**2**	7	7	7	7	**4**	**7**
A2	6	6	6	6	**8**	**6**	6	6	6	5	**9**	**6**	6	6	6	6	**8**	**6**
A3	4	4	4	4	**16**	**4**	4	3	1	2	**22**	**3**	4	4	2	3	**19**	**3**
A4	5	5	5	5	**12**	**5**	7	7	5	6	**7**	**7**	5	5	5	5	**12**	**5**
A5	1	1	3	1	**26**	**1**	2	2	2	3	**23**	**1**	1	1	1	1	**28**	**1**
A6	3	3	2	3	**21**	**3**	3	4	3	4	**18**	**4**	2	2	3	2	**23**	**2**
A7	2	2	1	2	**25**	**2**	5	5	4	7	**11**	**5**	3	3	4	4	**18**	**4**
Best Alt.	A5						A5						A5					

Note: The values of UDS and UDR (under UDM) and achieved alternatives are shown in bold text.

4.11. Case 3 under Optimal Load Growth: Assets (REG + D-STATCOM) Placements Equal to 0.90 LPF (C3_OLG)

4.11.1. Initial Evaluation of Alternatives in Case 3 under Optimal Normal Load (C3_OLG)

The initial evaluation of C3_OLG for each alternative in terms of TPE and CPE under OLG is shown in Table 25.

Table 25. Techno-economic evaluation analysis in case 3 (C3_OLG) for 33-bus MDN.

S#:		Technical Parameters Evaluations (TPE)							Cost (Economics Related) Parameters Evaluations (CPE)					
Case (No.) / Alt. (No).	DG Size (KVA) @ Bus Loc.	P_{Loss} (KW)	Q_{Loss} (KVAR)	*PLM* (%)	*QLM* (%)	*DGPP* (%)	V_{Min} (P.U)	P_{SSR} + j Q_{SSR} (KW) + j(KVAR)	*PLC* (M$)	*PLS* (M$)	*CPDG* ($/MWh)	*CQDG* ($/MVArh)	*AIC* (M$)	*ACD* (M$)
C3/A1	S1: 2187 + j1057 @15	179.6	105.58	60.14	65.4	38.74	0.9594	3325.6 + j2350.58	0.0944	0.9102	43.99	4.3573	0.4131	0.013941
C3/A2	S1: 3558 + j1723 @30	99.2	60.67	77.99	80.12	66.03	0.9656	1874.2 + j1639.67	0.0521	0.9524	71.35	7.1852	0.6721	0.022713
C3/A3	S1: 1269 + j622 @ 15 S2: 2223 + j1080 @ 30	58.83	34.39	86.95	88.73	66.23	0.9813	1899.83 + j1634.39	0.0309	0.9737	70.09	7.081	0.6596	0.022449
C3/A4	S1: 3150 + j1525 @ 30 S2: 531 + j257.1 @ 25	88.53	55.13	80.36	81.93	65.19	0.9657	1740.53 + j1575.03	0.0465	0.9581	73.87	7.3523	0.6953	0.023497
C3/A5	S1: 882.73 + j427 @15 S2: 2011 + j974 @ 30 S3: 1369 + j663.02 @ 7	36.26	22.37	91.95	92.66	77.99	0.9898	1106.53 + j1260.35	0.0191	0.9855	85.53	8.5578	0.8052	0.027276
C3/A6	S1: 1032 + j500@ 15 S2: 1907 + j923.8 @30 S3: 1145 + j554.5 @25	40.37	25.04	91.04	91.79	81.84	0.9842	1289.37 + j1348.74	0.0212	0.9834	81.94	8.1617	0.7714	0.026075
C3/A7	S1: 2601 + j1260 @ 30 S2: 531 + j257.1 @ 25 S3: 981 + j475.1 @ 8	47.03	28.63	89.56	89.56	72.92	0.9790	1267.03 + j1338.43	0.0247	0.9799	82.51	8.3215	0.7769	0.026255

4.11.2. MCDM Evaluation of Alternatives in Case 3 Under Optimal Load Growth (C3_OLG)

The MCDM evaluations for C3_OLG are illustrated in Figure 15a–d. Refer to Table 25 for respective TPE and CPE in terms of numerical details without normalization. The rank of the alternatives in C3_OLG is shown in Table 26. As per the results, the best alternative in TPE is A5, whereas there are no unanimous solutions in CPE and OPE. After applying UDM, the best solutions in TPE, CPE, and OPE are A5 (UDS=28 and UDR=1), A3 (UDS=23 and UDR=1), and A5 (UDS=26 and UDR=1); respectively. The UDM with UDS and UDR across TPE, CPE and OPE in C3_OLG have highlighted in bold text as shown in Table 26.

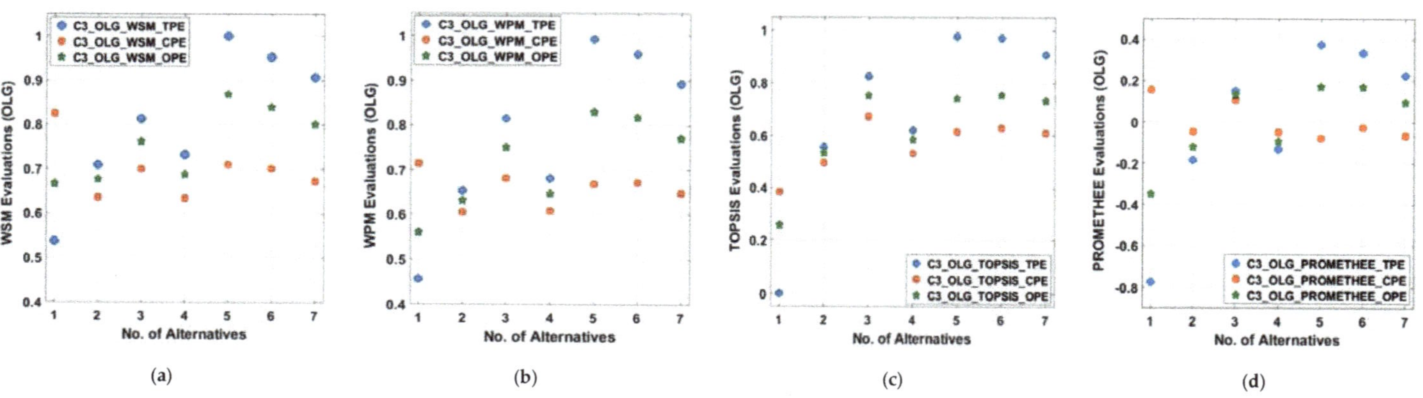

(a) (b) (c) (d)

Figure 15. MCDM evaluations for C3_OLG in 33-bus MDN: (**a**) WSM scores; (**b**) WPM scores; (**c**) TOPSIS scores; (**d**) PROMETHEE scores.

Table 26. Order of the ranks across TPE, CPE, and OPE in C3_OLG for 33-bus MDN.

Evaluations	TPE (C3_OLG)						CPE (C3_OLG)						OPE (C3_OLG)					
C3_OLG / Alt (#)	WSM	WPM	TOPSIS	PROMETHEE	UDS	UDR	WSM	WPM	TOPSIS	PROMETHEE	UDS	UDR	WSM	WPM	TOPSIS	PROMETHEE	UDS	UDR
A1	7	7	7	7	**4**	**7**	1	1	7	1	**22**	**2**	7	7	7	7	**4**	**7**
A2	6	6	6	6	**8**	**6**	6	7	6	4	**9**	**7**	6	6	6	6	**8**	**6**
A3	4	4	4	4	**16**	**4**	4	2	1	2	**23**	**1**	4	4	2	3	**19**	**3**
A4	5	5	5	5	**12**	**5**	7	6	5	5	**9**	**6**	5	5	5	5	**12**	**5**
A5	1	1	1	1	**28**	**1**	2	4	3	7	**16**	**4**	1	1	3	1	**26**	**1**
A6	2	2	2	2	**24**	**2**	3	3	2	3	**21**	**3**	2	2	1	2	**25**	**2**
A7	3	3	3	3	**20**	**3**	5	5	4	6	**12**	**5**	3	3	4	4	**18**	**4**
Best Alt.	A5						A3						A5					

Note: The values of UDS and UDR (under UDM) and achieved alternatives are shown in bold text.

4.12. Case 4 Under Optimal Load Growth: Asset REG + D-STATCOM Placements Equal to 0.85 LPF (C4_OLG)

4.12.1. Initial Evaluation of Alternatives in Case 4 under Optimal Load Growth (C4_OLG)

The initial evaluation of case 4 (C4_OLG) for each alternative under LG in terms of TPE and CPE is shown in Table 27.

Table 27. Techno-economic evaluation analysis in case 4 (C4_OLG) for 33-bus MDN.

S#:		Technical Parameters Evaluations (TPE)							Cost (Economics Related) Parameters Evaluations (CPE)					
Case (No.)/Alt. (No).	DG Size (KVA) @ Bus Loc.	P_{Loss} (KW)	Q_{Loss} (KVAR)	*PLM* (%)	*QLM* (%)	*DGPP* (%)	V_{Min} (P.U)	P_{SSR} + j Q_{SSR} (KW) + j(KVAR)	*PLC* (M$)	*PLS* (M$)	*CPDG* ($/MWh)	*CQDG* ($/MVArh)	*AIC* (M$)	*ACD* (M$)
C4/A1	S1: 2048 + j1270 @ 15	180.28	119.79	59.99	60.74	38.43	0.9594	3465.28 + j2151.79	0.0947	0.9098	41.22	6.4801	0.3868	0.01675
C4/A2	S1: 3336 + j2067 @ 30	95.95	66.98	78.71	78.05	62.56	0.9656	2092.95 + j1301.98	0.0504	0.9542	66.97	10.539	0.6301	0.026728
C4/A3	S1: 1181 + j739.3 @15 S2: 2082 + j1298@ 30	56.42	37.02	87.48	87.87	61.34	0.9814	2126.42 + j1301.72	0.0296	0.9749	65.53	10.476	0.6163	0.02687
C4/A4	S1: 2401 + j1488 @30 S2: 1275 + j790 @ 25	77.68	55.32	82.76	81.87	63.01	0.9666	1734.68 + j1079.32	0.0408	0.9638	73.77	11.701	0.6944	0.030028
C4/A5	S1: 833.7 + j516 @15 S2: 1899 + j1177 @ 30 S3: 1000 + j620.2 @ 7	36.64	24.02	91.87	92.13	81.82	0.9877	1636.94 + j1012.82	0.0193	0.9853	74.94	11.759	0.7051	0.030473
C4/A6	S1: 975.3 + j604.5 @15 S2: 1765 + j1094@ 30 S3: 1085.3 + j663 @25	37.69	26.11	91.64	91.44	80.34	0.9844	1545.1 + j966.61	0.0198	0.9848	76.76	11.914	0.7226	0.031107
C4/A7	S1: 1785 + j1106 @ 30 S2: 1190 + j737 @ 25 S3: 1156 + j716.4 @ 8	37.58	25.66	91.66	91.66	77.51	0.9825	1239.58 + j768.26	0.0197	0.9848	82.87	13.099	0.7803	0.033743

4.12.2. MCDM Evaluation of Alternatives in Case 4 under Optimal Load growth (C4_OLG)

The MCDM Evaluations for Case 4 (C4_OLG) under OLG Are Illustrated in Figure 16a–d. The rank of the alternatives in C4_OLG is shown in Table 28. As per the results, the best alternative in TPE and CPE is A5, whereas there are no unanimous solutions in CPE. After applying UDM, the best solutions in TPE, CPE, and OPE are A5 (UDS=28 and UDR=1), A5 (UDS=23 and UDR=1), and A5 (UDS=28 and UDR=1); respectively. The UDM with UDS and UDR across TPE, CPE and OPE in C4_OLG have highlighted in bold text as shown in Table 28.

In all the cases of the proposed IDMP approach above, it is found that the TPE of each respective case has less variance compared to other cases. The OPE shows the maximum variance while the CPE shows the maximum variance when evaluated across various MCDM methodologies.

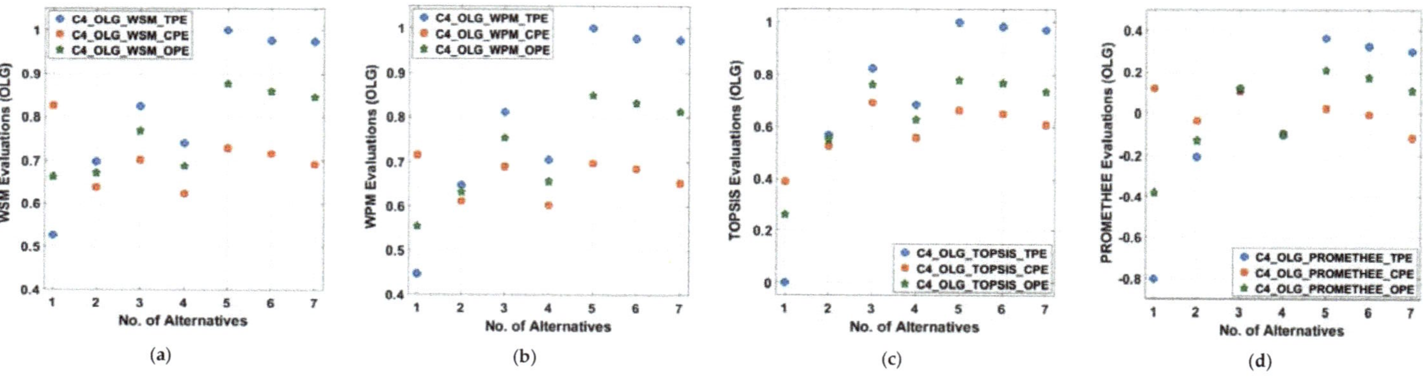

(a) (b) (c) (d)

Figure 16. MCDM evaluations for C4_OLG in 33-bus MDN: (**a**) WSM scores; (**b**) WPM scores; (**c**) TOPSIS scores; (**d**) PROMETHEE scores.

Table 28. Order of the ranks across TPE, CPE, and OPE in C4_OLG for 33-bus MDN.

Evaluations	TPE (C4_OLG)						CPE (C4_OLG)						OPE (C4_OLG)					
C4_OLG / Alt (#)	WSM	WPM	TOPSIS	PROMETHEE	UDS	UDR	WSM	WPM	TOPSIS	PROMETHEE	UDS	UDR	WSM	WPM	TOPSIS	PROMETHEE	UDS	UDR
A1	7	7	7	7	**4**	**7**	1	1	7	1	**22**	**3**	7	7	7	7	**4**	**7**
A2	6	6	6	6	**8**	**6**	6	6	6	5	**9**	**6**	6	6	6	6	**8**	**6**
A3	4	4	4	4	**16**	**4**	4	3	1	2	**22**	**2**	4	4	3	3	**18**	**4**
A4	5	5	5	5	**12**	**5**	7	7	5	6	**7**	**7**	5	5	5	5	**12**	**5**
A5	1	1	1	1	**28**	**1**	2	2	2	3	**23**	**1**	1	1	1	1	**28**	**1**
A6	2	2	2	2	**24**	**2**	3	4	3	4	**18**	**4**	2	2	2	2	**24**	**2**
A7	3	3	3	3	**20**	**3**	5	5	4	7	**11**	**5**	3	3	4	4	**18**	**3**
Best Alt.			**A5**						**A5**						**A5**			

Note: The values of UDS and UDR (under UDM) and achieved alternatives are shown in bold text.

5. Comparison and Validation Analysis

The proposed IDMP approach aimed at (multiple-loop configured) MDN is evaluated on 33-bus TDN and validated via comparative analysis of achieved results with the findings in the available literature, respectively. The comparison section consists of two sub-sections. In the first section, achieved results are compared with each other along with the change of rank in different cases of asset sitting and sizing. In the following case, the achieved results are compared with the results reported in the reviewed literature.

5.1. Results Comparison with Achieved Results

The achieved results for self-comparison are presented in Table 29. In Table 29, an overview of performance evaluation in the proposed IDMP approach across technical, cost, and overall (techno-economic) criteria were evaluated under various load and generation conditions in 33-bus TDS.

In case 1 (C1), from the perspective of technical evaluation, the TPE across NL, LG, and OLG as designated by TPE_NL, TPE_LG, and TPE_OLG are presented after applying UDM. The relative comparison from the viewpoint of rank changed (RC) designated by RC1, between respective cases of NL and LG, reveals that in C1, all ranks are changed under TPE except A7. The RC2 among NL and OLG reveals the same. However, in RC3, the only ranks changed amongst solutions are A1–A3.

In C1, from the perspective of cost-economic evaluation, the CPE across NL, LG, and OLG as designated by CPE_NL, CPE_LG, and CPE_OLG are presented after applying UDM. The results indicate that in RC1 and RC2, all the ranks of possible solutions (alternatives) have changed. However, in RC3, the ranks changed in the achieved solutions are A1–A4, respectively.

In C1, from the perspective of overall (techno-economic) performance evaluation, the OPE across NL, LG, and OLG as designated by OPE_NL, OPE_LG, and OPE_OLG are presented after applying UDM. The achieved results indicate that in RC1 and RC2, all the ranks of alternatives have changed except A7. However, in RC3, the ranks changed in the achieved solutions are A1–A2.

In C2, in terms of TPE, RC1 shows the rank change in A5, A6, and A7. In RC2, the change of ranks is found in A1–A2, A4–A6. In RC3, the change of rank is found in A1 and A2. In terms of CPE, in RC1, change of ranks is observed in A1–A2. In RC2, ranks change is observed in all alternatives. In RC3, change of rank is observed in A1–A2 and A4. From the viewpoint of OPE, in RC1, change of rank is observed in all alternatives except A7. In RC2, rank change is observed in all except A2 and A7. In RC3, rank change is observed in solutions designated by A2–A4.

It is also observed that in C1–C2, DG can be subjected to reactive power support limit whereas in C3–C4, when REG and D-STATCOM are decoupled, overall better performance is achieved as mentioned throughout the paper, as demonstrated in the results and discussion section.

In C3, in terms of TPE, RC1 shows the rank change in A1–A3. In RC2, no change of ranks is found. In RC3, the change of rank is found in A1–A3. In terms of CPE, in RC1, change of ranks is observed in A1–A4. In RC2, ranks change is observed in all alternatives except A3 and A5. In RC3, change of rank is observed in A3–A4 and A6–A7. From the viewpoint of OPE, in RC1, change of rank is observed in all alternatives except A1–A2. In RC2, in respective comparison, all ranks changed. In RC3, rank change is observed in solutions designated by A1–A2.

In C4, in terms of TPE, RC1 shows no change of rank amongst stated alternatives. In RC2, change of ranks is found in A1–A3. In RC3, the change of rank is found in A1–A3. In terms of CPE, in RC1, change of ranks is observed in A2–A4. In RC2, ranks change is observed in alternatives A1–A4. In RC3, change of rank is observed in A1–A3. From the viewpoint of OPE, in RC1, change of rank is observed in A2–A3. In RC2, in respective comparison, no ranks have changed. In RC3, rank change is observed in solutions designated by A2–A3.

Table 29. Overview performance evaluation analysis for all evaluated cases of IDMP across various evaluations for self-comparison of change in ranks.

C#/A#:		Technical Evaluation						Cost-Economic Evaluation						Techno-Economic Evaluation					
Cases #:	Alternatives #:	TPE_NL	TPE_LG	TPE_OLG	RC1: NL-LG	RC2: NL-OLG	RC3: LG-OLG	CPE_NL	CPE_LG	CPE_OLG	RC1: NL-LG	RC2: NL-OLG	RC3: LG-OLG	OPE_NL	OPE_NL	OPE_NL	RC1: NL-LG	RC2: NL-OLG	RC3: LG-OLG
	A1	7	7	7				4	1	2				7	7	7			
	A2	5	6	6				2	7	7				5	6	6			
	A3	6	4	4				7	2	1				6	3	3			
Case-1	A4	3	5	5	**All except A7**	**All except A7**	**A1, A2, A3**	1	6	6	**All**	**All**	**A1, A2, A3, A4**	3	5	5	**All except A7**	**All except A7**	**A1, A2**
	A5	2	3	1				6	3	4				2	2	1			
	A6	4	1	2				5	4	3				4	1	2			
	A7	1	2	3				3	5	5				1	4	4			
	A1	7	7	7				5	3	3				7	7	7			
	A2	5	6	6				3	6	6				5	6	6			
	A3	6	4	4				1	1	2				6	4	3			
Case-2	A4	4	5	5	**A5, A6, A7**	**A1, A2, A4, A5, A6**	**A1, A2**	6	7	7	**A1, A2**	**All**	**A1, A2, A4**	4	5	5	**All except A7**	**A1, A3-A6**	**A2, A3, A4**
	A5	2	2	1				7	4	1				3	1	1			
	A6	3	3	3				2	2	4				2	3	2			
	A7	1	1	2				4	5	5				1	2	4			
	A1	7	7	7				4	2	2				7	7	7			
	A2	6	6	6				6	6	7				6	6	6			
	A3	4	4	4				2	1	1				3	3	3			
Case-3	A4	5	5	5	**A1, A2, A3**	**Nil**	**A1, A2, A3**	7	7	6	**A1, A2, A3, A4**	**A1, A2, A4,A6,A7**	**A3, A4, A6, A7**	5	5	5	**A1, A2**	**All**	**A1, A2**
	A5	1	3	1				1	3	4				1	2	1			
	A6	2	1	2				3	4	3				2	1	2			
	A7	3	2	3				5	5	5				4	4	4			
	A1	7	7	7				4	2	3				7	7	7			
	A2	6	6	6				6	6	6				6	6	6			
	A3	4	4	4				1	1	2				4	4	4			
Case-4	A4	5	5	5	**Nil**	**A1, A2, A3**	**A1, A2, A3**	7	7	7	**A2, A3, A4**	**A1, A2, A3, A4**	**A1, A2, A3**	5	5	5	**A2, A3**	**Nil**	**A2, A3**
	A5	2	2	1				2	3	1				1	1	1			
	A6	3	3	2				3	4	4				2	3	2			
	A7	1	1	3				5	5	5				3	2	3			

Note: RC: Rank changed; RC1: Rank changed among cases of NL-LG; RC2: Rank changed among cases of NL-OLG; RC2: Rank changed among cases of LG-OLG (shown in bold text).

5.2. Results Comparison with Reported Results

The proposed IDMP approach aims at MDN is evaluated on the 33-bus TDS and validated via comparative analysis of achieved results with the findings in the available literature, respectively. The achieved results are compared on the basis of best-achieved solutions obtained in each case (C1–C4) via proposed approach across NL, LG, and OLG via each TPE, CPE, and OPE, respectively.

5.2.1. Evaluated Results Comparison of C1_NL for DGs Operating at 0.90 LPF

The evaluation comparison of case 1 (C1_NL) for each alternative in terms of TPE, CPE, and OPE are shown in Table 30. The achieved results are compared with multi-objective hybrid GA and TOPSIS approach in [42] and the multi-objective centric hybrid sensitivity-based approach in [53]. It is observed that the proposed alternative A7 in IDMP during C1_NL is best from the viewpoint of TPE and OPE, whereas A4 outperforms on the basis of CPE. Note that the achieved results that outperformed the compared works are shown in bold text, throughout this section.

Table 30. Comparisons of results with C1_NL for 33-bus TDN (DG@LPF = 0.90).

Performance Evaluation Indicators (PEIs)	[42]	[42]	[53]	A4 [31,31] TPE (C1_NL) OPE (C1_NL)	
DG Size (KVA) @DG Site (Bus)	773 @ 14 378 @ 25 847 @ 30	700 @ 15 430 @ 18 870 @ 28	2074.56@6 615.25@15	540@25 2357@30	**1957@30** **500 @25** **760@8**
P_{Loss} (KW)	28.83	39.76	65.8435	32.99	**18.870**
Q_{Loss} (KVAR)	-	-	51.94	25.491	**13.327**
PLM (%)	86.33	81.15	68.8	84.37	**91.06**
QLM (%)	-	-	63.7	82.17	**90.68**
DG Capacity (KVA)	1998	2000	2689.81	2897	**3217**
DGPP (%)	45.73	45.77	61.56	66.303	**73.63**
P_{SSR} + j Q_{SSR}	-	-	1347.9 + j 836.34	1140.7 + j 1062.72	**838.570 + j911.07**
V_{Min} (P.U)	0.9756@30	0.9796@25	0.97567	0.9773@13	**0.9857@14**
PLC (Million-$)	-	-	0.03461	0.01261	**0.00992**
PLS (Million-$)	-	-	0.07629	0.09829	**0.1010**
CPDG ($/MWh)	-	-	-	**52.396**	58.156
CQDG ($/MVArh)	-	-	-	**5.2141**	5.7938
AIC (Million-$)	-	-	-	**0.5235**	0.5813

Note: The outperformed results in comparative study are shown in bold text.

5.2.2. Evaluated Results Comparison of C2_NL for DGs Operating at 0.85 LPF

The evaluation comparison of case 2 (C2_NL) for each alternative under NL is shown in terms of TPE, CPE, and OPE are shown in Table 31. The best-achieved alternatives are compared with the multi-objective hybrid GA and TOPSIS approach in [42], loss sensitivity factor (LSF) and simulated annealing (SA)-based hybrid method in [54], heuristic-based krill herd algorithm in [55], and ant colony optimization (ACO) and artificial bee colony (ABC) as reported in [56], respectively. It is worth mentioning that the reported studies are more focused on technical evaluation, and comparison with CPE in the reported work is only presented for reference. It is observed that the proposed alternative A7 in IDMP during C2_NL is best from the viewpoint of TPE and OPE, whereas A3 outperforms on the basis of CPE. The outperformed results have shown in bold text for comparative analysis.

Table 31. Comparisons of results with C2_NL for 33-bus TDN (DG@LPF = 0.85).

Performance Evaluation Indicators (PEIs)	[42]	[54]	[55]	[56]	A3 [30] CPE (C2_NL)	A7 [31] TPE (C2_NL) OPE (C2_NL)
DG Size (KVA) @DG Site (Bus)	807@8 347@17 845@30	1382@6 550@18 1062@30	853@13 900@24 899@30	1014@12 960@25 1363@30	950@15 1633@30	**1422.1@30** **1045.4 @25** **933.4@8**
P_{Loss} **(KW)**	24.98	26.72	19.57	15.91	38.3	**13.85**
Q_{Loss} **(KVAR)**	-	-	-	-	28.1	**11.50**
PLM (%)	88.16	87.34	90.725	92.46	81.85	**94.44**
QLM (%)	-	-	-	-	80.35	**91.96**
DG Capacity (KVA)	1999	2994	2652	2880	2583	**3400.9**
DGPP (%)	45.75	68.523	60.70	65.91	59.12	**77.834**
P_{SSR} + j Q_{SSR}	-	-	-	-	1557.8 + j 967.42	**838.085 + j 519.965**
V_{Min} **(P.U)**	-	-	-	-	0.9719	**0.9880@15**
PLC (Million-$)	-	-	-	-	0.0201	**0.00728**
PLS (Million-$)	-	-	-	-	0.0908	**0.10362**
CPDG ($/MWh)	-	-	-	-	**44.161**	58.0651
CQDG ($/MVArh)	-	-	-	-	**6.94**	9.1375
AIC (Million-$)	-	-	-	-	**0.46673**	0.6145

Note: The outperformed results in comparative study are shown in bold text.

5.2.3. Evaluated Results Comparison of C3_NL and C4_NL for REG + D-STATCOM

The evaluation comparison of C3_NL and C4_NL for each alternative in terms of TPE, CPE, and OPE are shown in Tables 32 and 33 respectively.

The best-achieved alternatives in C3_NL are compared in Table 18 with well-established approaches such as the best-achieved alternatives and compared with well-established methods such as hybrid fuzzy ant colony optimization approach in [34], multiple attribute decision-making (MCDM) methods such as TOPSIS and PROMETHEE in [35], and sensitivity-based approach in [57]. It is found that A3 in [30] amongst other alternatives in the C3_NL of the proposed IDMP method provides a big picture on the basis of CPE. Moreover, on the basis of TPE and OPE, the findings of the A7 solution are in close agreement with the reported works, hence validating the proposed approach under NL.

The findings of C4_NL are compared in Table 18 with reported works such as hybrid fuzzy ant colony optimization method in [34] and cuckoo search algorithm (CSA) in [37]. The reported work is in close agreement with solution A7 in [37] on the basis of TPE, A3 [30] and A5 [30] on the basis of CPE and OPE, which indicates the validity of proposed approach with assets such as REG and DSTATCOM, respectively.

Table 32. Comparisons of results with C3_NL for 33-bus TDN (REG + D-STATCOM@LPF = 0.9).

Performance Evaluation Indicators (PEIs)	[34]	[57]	[35]	A3 [30] CPE (C3_NL)	A5 [30] TPE (C3_NL) OPE (C3_NL)
DG (KW) @ Bus # D-STATCOM (KVAR) @ Bus #	1316@9 740@10	2491 @ 6 1230@30	750 @ 14 420 @ 14 1100 @ 24 460 @ 24 1000 @ 8 970 @ 8	869.2 @ 15 421.2 @ 15 1604 @ 30 777.04 @ 30	620.5 @ 15 300 @ 15 1442 @ 30 698.3 @ 30 637.5 @ 7 308.73 @ 7
P_{Loss} **(KW)**	48.73	58	**15.07**	27.89	19.40
Q_{Loss} **(KVAR)**	-	-	-	16.20	**11.09**
PLM (%)	76.9	72.51	**92.56**	86.52	90.63

Table 32. *Cont.*

QLM (%)	-	-	-	88.44	**92.09**
DG Capacity (KW)	1316	2491	2460	2473.2	**2700**
D-STATCOM Capacity (KVAR)	740	1230	1600	1198.24	**1307.03**
DGPP (%)	34.56	67	67.2	67.56	**73.77**
$P_{SSR} + j\ Q_{SSR}$	-	-	-	1770+j 1118	**1034 + j 1004.1**
V_{Min} (P.U)	-	-	0.9584	0.9874	**0.9900**
PLC (Million-$)	-	-	-	0.0147	**0.0102**
PLS (Million-$)	-	-	-	0.0962	**0.1007**
CPDG ($/MWh)	-	50.1	-	**49.71**	54.25
CQDG ($/MVArh)	-	5.2	-	**4.9558**	5.3593
AIC (Million-$)	-	-	-	**0.4672**	0.5099
ACD (Million-$)	-	-	-	**0.01589**	0.01718

Note: The outperformed results in comparative study are shown in bold text.

Table 33. Comparisons of results with C4_NL for 33-bus TDN (REG + D-STATCOM@LPF = 0.85).

Performance Evaluation Indicators (PEIs)	[34]	[37]	A7 [31] TPE (C4_NL)	A3 [30] CPE (C4_NL)	A5 [30] OPE (C4_NL)
DG (KW) @ Bus # D-STATCOM (KVAR) @ Bus #	1309 @ 7 720 @ 23	850 @ 12 400 @ 12 750 @ 25 350 @ 25 860 @ 8 850 @ 8	1210 @ 30 750 @ 30 890 @ 25 551.2 @ 25 793.7 @ 8 492 @ 8	807.5 @ 15 485 @ 15 1388 @ 30 893 @ 30	547 @ 15 338.8 @ 15 1397 @ 30 866 @ 30 606.3 @ 7 376 @ 7
P_{Loss} (KW)	69.15	**12**	16.3	27.39	17.33
Q_{Loss} (KVAR)	-	-	12.6	17.98	**11.37**
PLM (%)	67.23	**94.31**	92.27	86.77	91.62
QLM (%)	-	-	91.19	87.18	**91.89**
DG Capacity (KW)	1309	2850	**2893.7**	2195.5	2550.3
D-STATCOM Capacity (KVAR)	720	1850	**1793.2**	1378	1580.8
DGPP (%)	34.19	76.72	**77.89**	60	68.67
$P_{SSR} + j\ Q_{SSR}$	-	-	**838.06 + j 519.4**	1547 + j940	1182 + j730.6
V_{Min} (P.U)	-	0.9862	0.9878	0.9859	**0.9901**
PLC (Million-$)	-	-	**0.0086**	0.0144	0.0091
PLS (Million-$)	-	-	**0.1023**	0.0965	0.1018
CPDG ($/MWh)	-	-	58.07	**44.16**	51.25
CQDG ($/MVArh)	-	-	9.1760	**7.1574**	8.0011
AIC (Million-$)	-	-	0.64653	**0.4147**	0.4817
ACD (Million-$)	-	-	0.02368	**0.01821**	0.02077

Note: The outperformed results in comparative study are shown in bold text.

5.2.4. Evaluated Results Comparison of C1_LG-C4_LG

The evaluation comparison of all four cases under load growth is compared with the results of the sensitivity-based approach reported in [57]. The results are compared for the cases C1_LG and C2_LG in Table 34 for the assets considering DG only case operating at 0.90 LPF and 0.85 LPF, respectively. The assets in both cases, such as DGs only, are capable of contributing both active and reactive power. The case for LG is presented for the reason that the LG impact from the perspective of assets considered has to be evaluated and relative impacts from the techno-economic perspective must be assessed. From the viewpoint of C1_LG, it is found that solution A3 via CPE and A6 via TPE and OPE is the best that even outperforms the results in [57]. From the standpoint of C2_LG, A7 (in TPE), A3 (in CPE), and A5 in OPE outperform the results stated in [57].

Table 34. Comparisons of results with C1_LG and C2_LG for 33-bus TDN (DG@LPF = 0.90, 0.85).

Performance Evaluation Indicators (PEIs)	[57]	A3 CPE (C1_LG)	A6 TPE, OPE (C1_LG)	A7 TPE (C2_LG)	A3 CPE (C2_LG)	A5 OPE (C2_LG)
DG Size (KVA) @DG Site (Bus)	4441 @ 6	971 @ 15 1783 @ 30	894.6 @ 15 1386 @ 30 822.6 @ 25	1422 @ 30 1045 @ 25 933.4 @ 8	950 @ 15 1633 @ 30	**828.3 @ 15** **1644 @ 30** **727.8 @ 7**
P_{Loss} (KW)	146	95.35	**77.621**	**71.17**	101.56	81.44
Q_{Loss} (KVAR)	-	53.664	**44.204**	**46.64**	65	51.93
PLM (%)	68.94	78.842	**82.776**	**84.21**	77.44	81.93
QLM (%)	-	82.42	**85.51**	**84.72**	78.69	82.98
DG Capacity (KVA)	4441	2754	3103.2	**3400.4**	2583	3200.1
DGPP (%)	**70.81**	43.802	49.131	54.22	41.18	47.83
P_{SSR} + j Q_{SSR}	-	2950 + j2155	2618 + j1994	2514 + j1557	3239 + j2006	**2694 + j1669**
V_{Min} (P.U)	-	0.9641	0.9652	0.9597	0.9624	**0.9641**
PLC (Million-$)	-	0.1762	0.1361	**0.1243**	0.1792	0.1349
PLS (Million-$)	-	0.8284	0.8684	0.8802	0.8254	**0.8697**
CPDG ($/MWh)	73.6	48.80	55.77	58.07	**44.16**	51.26
CQDG ($/MVArh)	13.7	4.8567	5.4988	9.1375	**7.1279**	8.0612
AIC (Million-$)	-	0.4965	0.5569	0.6145	**0.4667**	0.5421

Note: The outperformed results in comparative study are shown in bold text.

The results are compared for the cases C3_LG and C4_LG in Table 35 for the assets such as REG and D-STATCOM, which are decoupled in comparison with DG only cases reported in C1_LG and C2_LG. However, they contribute active and reactive power that is equal to one DG supplying P and Q either at 0.90 or 0.85 LPF. The P contributes via REG and Q support is provided by D-STATCOM. It is found that the achieved results outperform reported works on the basis of TPE, CPE, and OPE.

Table 35. Comparisons of results with C3_LG and C4_LG for 33-bus TDN (REG + D-STATCOM).

Performance Evaluation Indicators (PEIs)	[57]	A3 CPE (C3_LG)	A6 TPE, OPE (C3_LG)	A7 TPE (C4_LG)	A3 CPE (C4_LG)	A5 OPE (C4_LG)
DG (KW) @ Bus # D-STATCOM (KVAR) @ Bus #	3670 @ 6 1770@30	869.2 @ 15 421.5 @ 15 1604 @ 30 777.4 @ 30	789 @ 15 380.7 @ 15 1247 @ 30 586.2 @ 30 739.6 @ 25 372 @ 25	1210 @ 30 750 @ 30 890 @ 25 551.2 @ 25 793.7 @ 8 492 @ 8	807.5 @ 15 485.0 @ 15 1388 @ 30 893 @ 30	547 @ 15 338.8 @ 15 1397 @ 30 866 @ 30 606.3 @ 7 376 @ 7
P_{Loss} (KW)	126	96.74	**78.98**	**72.69**	102.66	82.85
Q_{Loss} (KVAR)	-	54.28	**44.87**	**47.27**	65.38	53.07
PLM (%)	73.19	78.53	**82.47**	**83.87**	77.22	81.69
QLM (%)	-	82.21	**85.29**	**84.51**	78.58	82.61
DG Capacity (KVA)	3670	2473.2	2775.6	**2893.7**	2195.5	2550.3
DGPP (%)	**68.82**	43.81	49.21	54.22	41.39	47.82
P_{SSR} + j Q_{SSR}	-	2957 + j 2158	2636+j2008	2512+j1556	**3240 + j1989**	2866 + j1774
V_{Min} (P.U)	-	0.9640	**0.9651**	0.9595	0.9623	0.9638

Table 35. *Cont.*

PLC (Million-$)	-	0.1789	0.1398	**0.1289**	0.1848	0.1373
PLS (Million-$)	-	0.8256	0.8648	**0.8757**	0.8198	0.8673
CPDG ($/MWh)	73.6	49.71	55.76	58.07	**44.16**	51.25
CQDG ($/MVArh)	7.3	**4.948**	5.6037	9.145	7.2153	8.0446
AIC (Million-$)	-	0.4672	0.5243	0.5466	**0.4147**	0.4817
ACD (Million-$)	-	**0.01579**	0.01781	0.023616	0.01828	0.020826

Note: The outperformed results in comparative study are shown in bold text.

5.2.5. Evaluated Results Comparison of C1_OLG-C4_OLG

The evaluation comparison of achieved results in C1_OLG and C2_OLG are shown in Table 36, the multi-aspect results outperform the reported results in [57].

Table 36. Comparisons of results with C1-C2 under OLG for 33-bus TDS (DG@LPF = 0.90, 0.85).

Performance Evaluation Indicators (PEIs)	[57]	A3, CPE (C1_OLG)	A5, TPE, OPE (C1_OLG)	A5, TPE, CPE, OPE (C2_OLG)
DG Size (KVA) @DG Site (Bus)	4441 @ 6	1500 @ 15 2300 @ 30	**980 @ 15** **2235 @ 30** **1521 @ 7**	**980 @ 15** **2235 @ 30** **1177 @ 7**
P_{Loss} **(KW)**	146	60.34	34.99	**34.63**
Q_{Loss} **(KVAR)**	-	35.29	**21.80**	23.32
PLM (%)	68.94	86.62	92.23	**92.31**
QLM (%)	-	88.43	92.86	**92.36**
DG Capacity (KVA)	4441	3800	**4736**	4392
DGPP (%)	70.81	60.58	**75.52**	70.04
$P_{SSR} + j\ Q_{SSR}$	-	**1973 + j1681**	1106 + j1259	1634 + j1012
V_{Min} **(P.U)**	-	0.9811	0.9899	**0.9878**
PLC (Million-$)	-	0.0317	0.0184	**0.0182**
PLS (Million-$)	-	0.9729	0.9862	**0.9864**
CPDG ($/MWh)	73.6	**68.65**	85.51	74.94
CQDG ($/MVArh)	13.7	**6.8440**	8.5309	11.804
AIC (Million-$)	-	**0.6867**	0.8559	0.7938

Note: The outperformed results in comparative study are shown in bold text.

The evaluation results of C3_OLG and C4_OLG are compared in Table 37 with hybrid particle swarm optimization (PSO) and GAMS in [36], and sensitivity-based approach in [57]. The results from the proposed work outperforms the reported works from the perspective of better performance and optimal sizing assets, as shown in bold text shown throughout this section.

Table 37. Comparisons of results with C4-C4 under OLG for 33-bus TDS (REG + D-STATCOM).

Performance Evaluation Indicators (PEIs)	[57]	[36]	[36]	A3 CPE (C3_OLG)	A5 TPE, OPE (C3_OLG)	A5 OPE (C4_OLG)
DG (KW) @ Bus # D-STATCOM (KVAR) @ Bus #	3670 @ 6 1770@30	1777.2@30 1123.5 @ 8	1829.7@30 689.3 @ 8	1269 @ 15 622.0 @ 15 2223 @ 30 1080 @ 30	882.73 @ 15 427 @ 15 2011 @ 30 974 @ 30 1369 @ 25 663.02 @ 25	833.7 @ 15 516 @ 15 1899 @ 30 1177 @ 30 1000 @ 7 620.2 @ 7
P_{Loss} (KW)	126	233.73	235.7	58.83	**36.26**	36.64
Q_{Loss} (KVAR)	-	-	-	34.39	**22.37**	24.02
PLM (%)	73.19	30.35	29.94	86.95	**91.95**	91.87
QLM (%)	-	-	-	88.73	**92.66**	92.13
DG Capacity (KVA)	3670	0	0	3492	**4262.73**	3732.7
D-STATCOM Size (KVAR)	1770	2900.7	2519	1702	2064.02	**2312.2**
DGPP (%)	68.82	0	0	66.23	77.99	**81.82**
P_{SSR} + j Q_{SSR}	-	-	-	1900+j1634	1107 + j1260	1367+ j1013
V_{Min} (P.U)	-	0.9447	0.94305	0.9813	**0.9898**	0.9877
PLC (Million-$)	-	0.122849	0.123884	0.0309	**0.0191**	0.0193
PLS (Million-$)	-	0.038616	0.0396	0.9737	**0.9855**	0.9853
CPDG ($/MWh)	73.6	-	-	**70.09**	85.53	74.94
CQDG ($/MVArh)	7.3	-	-	**7.081**	8.5578	11.759
AIC (Million-$)	-	**0.161464**	0.16349	0.6596	0.8052	0.7051
ACD (Million-$)	-	0.015374	**0.01335**	0.02245	0.02728	0.030473

Note: The outperformed results in comparative study are shown in bold text.

6. Conclusions

The active meshed distribution network is considered as a model of future smart distribution networks, which are anticipated to exhibit better performance and reliability through interconnection. Practical planning problems must be capable of encapsulating solutions that must satisfy conflicting criteria across the planning horizon. Thus, this works offers an IDMP approach aimed at various types of asset sitting and sizing across the normal load and load growth levels in a meshed distribution network. The assets considered in this study are synchronous generators operating at various lagging power factors (LPF) and capable of giving active and reactive powers and renewable DGs like photovoltaic (PV) system (contributes active power only) and D-STATCOM (contributes reactive power only). The methodology consists of an initial evaluation of alternatives with the voltage stability assessment indices-loss minimization condition (*VSAI-LMC*)-based method with single and multiple assets. Later, four MCDM methodologies are applied to sort out the best solution amongst the achieved alternatives. Finally, unanimous decision-making (UDM) is applied to find out one trade-off solution amongst achieved ranks of various MCDM methods. The methodology is applied across technical only (TPE), cost-economic only (CPE), and overall techno-economic (OPE) performance evaluations across four cases of assets sitting and sizing. All four cases are evaluated across the normal load, load growth, and optimal load growth. The detailed performance analysis is applied across the meshed configured 33-bus test distribution network. The achieved results across all cases after comparison with the credible results reported in the literature have both outperformed and displayed close agreement, resulting in the validation of the proposed IDMP approach. The proposed approach with techno-economic performance evaluation among conflicting criteria across the time scale reduces the need for sensitivity analysis and provides a range of trade-off solutions across various performance metrics. The overall approach in this paper aims to offer decision-makers a wide variety of optimal solutions among conflicting criteria considering various cases of asset optimization.

Supplementary Materials: The following are available online at http://www.mdpi.com/1996-1073/13/6/1444/s1.

Author Contributions: S.A.A.K. (First and corresponding author), U.A.K. (Second author), H.W.A. (Third Author), S.A. (Fourth Author) and D.R.S. (Fifth author) offers the presented study. The roles are defined as per the order of authors. S.A.A.K. is responsible for conceptualization, methodology, formal analysis, and investigation, resources, writing (original draft preparation), writing (review and editing), supervision, project administration, funding acquisition. U.A.K., H.W.A., and S.A. are responsible for formal analysis, investigation, software, validation, visualization, and writing (original draft preparation). D.R.S. is responsible for supervision and review. All authors have read and agreed to the published version of the manuscript

Funding: This research received no external funding. This research was supported and conducted in the United States-Pakistan Center of Advanced Studies in Energy (USPCAS-E), National University of Science and Technology (NUST), Islamabad, Pakistan. The APC is funded internally by National University of Science and Technology (NUST), Islamabad, Pakistan.

Conflicts of Interest: The authors declare no conflict of interest.

Abbreviations

The following abbreviations have used in this paper.

ACD	Annual cost of D-STATCOM
ADN	Active distribution Network
AIC	Annual investment cost
AF_c	Annualized factor (of cost) in USD $
C (#)	Case (No. = 1, 2, 3, 4)
C_t	Annual cost based on interest-rate
CPDG	Cost of active power from DG
CPE	Cost-economic performance evaluation
CQDG	Cost of reactive power from DG
CU_c	Cost related to DG unit (USD/KVA)
DG	Distributed generation units
DGPP	DG penetration by percentage in TDN
DM	Decision-making
DN	Distribution network
DNPP	Distribution network planning problems
D-STATCOM	Distributed static compensator
DS/DSt	D-STATCOM
DGC_{max}	Maximum capacities of DG units in (KVA)
Eqn. (No)	Equation. (Number)
E_U	Rate of electricity unit
GA	Genetic algorithm
IDMP	Integrated decision making planning
LDN	Loop distribution network
LG	Load growth
LM	Loss minimization
LMC	Loss minimization condition
LPF	Lagging power factor
LSF	Loss sensitivity factor
MCDM	Multi criteria decision making
MDN	Meshed distribution network
M$	Millions of USD ($)
NL	Normal load
NO	Normally open

NSGA-II	Non dominated GA-II
ODGP	Optimal DG placement
ODGP	Optimal DG Unit Placement
OLG	Optimal load growth
OPE	Overall (techno-economic) performance evaluation.
P	Active Power
PEI	Performance evaluation indicator
P_{ss}/P_{DG}	P contribution from substation & DG
PF/pf	Power factor
P_{Loss}	Active Power loss in KW
PLC	Cost of P_{Loss} (in million USD)
PLS	Active power loss saving in Million $
P_{SSR}	P Capacity Release from Substation
P.U	Per unit system values (or p.u)
PRO/PROMETHEE	Preference ranking organization method for enrichment of evaluation
PSO	Particle swarm optimization
PV	Photovoltaic systems
Q	Reactive Power
Q_{DG}	Q contribution from substation
Q_{Loss}	Reactive Power loss in KVAR
QLM	Q_{Loss} minimization (by percentage)
Q_{SSR}	Q Capacity Release from Substation
RB	Receiving end (load) bus
RC	Rank changed
RDN	Radial-structured distribution network
PLM	P_{Loss} minimization (by percentage)
REG	Renewable energy generation
RSS	Relief-in-substation (P and Q) capacity
RTUs	Remote terminal units
S (#)	Set (No. = 1, 2, 3, 4) of assets
SA	Simulated annealing
SB	Sending end (feeding) bus
SCC	Short circuit current
SG	Smart grid
SS	Substation
TB	Tie-line branch
TDN	Test distribution Network
TOP/TOPSIS	Technique for order preference by similarity to ideal solution
TPE	Technical performance evaluation
TS/TS#	Tie-Switch (normally open switch)/TS.No.
T_Y	Time in a year = 8760 Hours
U_Max	Voltage maximization
UDM	Unanimous decision making
UDR	Unanimous decision making rank
UDS	Unanimous decision making score
V	Voltage magnitude
Vmin	Minimum voltage magnitude
VM	Voltage maximization
VP/VS	Voltage profile/Voltage stabilization
VSI/*VSAI*	Voltage stability assessment indices
WPM	Weighted product method
WSM	Weighted sum method
V_A	Feasible voltage solution via *VSAI*_A
V_B	Feasible voltage solution via *VSAI*_B
A^+, A^-	Positive and Negative ideal solution

References

1. Keane, A.; Ochoa, L.; Borges, C.L.T.; Ault, G.W.; Alarcon-Rodriguez, A.; Currie, R.A.F.; Pilo, F.; Dent, C.; Harrison, G.P. State-of-the-Art Techniques and Challenges Ahead for Distributed Generation Planning and Optimization. *IEEE Trans. Power Syst.* **2012**, *28*, 1493–1502. [CrossRef]
2. Fang, X.; Misra, S.; Xue, G.; Yang, D. Smart Grid—The New and Improved Power Grid: A Survey. *IEEE Commun. Surv. Tutorials* **2011**, *14*, 944–980. [CrossRef]
3. Alarcon-Rodriguez, A.; Ault, G.; Galloway, S. Multi-objective planning of distributed energy resources: A review of the state-of-the-art. *Renew. Sustain. Energy Rev.* **2010**, *14*, 1353–1366. [CrossRef]
4. Evangelopoulos, V.A.; Georgilakis, P.; Hatziargyriou, N.D. Optimal operation of smart distribution networks: A review of models, methods and future research. *Electr. Power Syst. Res.* **2016**, *140*, 95–106. [CrossRef]
5. Kim, J.-C.; Cho, S.-M.; Shin, H.-S. Advanced Power Distribution System Configuration for Smart Grid. *IEEE Trans. Smart Grid* **2013**, *4*, 353–358. [CrossRef]
6. Valenzuela, A.; Inga, E.; Simani, S. Planning of a Resilient Underground Distribution Network Using Georeferenced Data. *Int. J. Energies* **2019**, *12*, 644. [CrossRef]
7. Prakash, P.; Khatod, D.K. Optimal sizing and sitting techniques for distributed generation in distribution systems: A review. *Renew. Sustain. Energy Rev.* **2016**, *57*, 111–130. [CrossRef]
8. Mahmoud, P.H.A.; Phung, D.H.; Vigna, K.R. A review of the optimal allocation of distributed generation: Objectives, constraints, methods, and algorithms. *Renew. Sustain. Energy Rev.* **2017**, *75*, 293–312.
9. Georgilakis, P.; Hatziargyriou, N.D. Optimal Distributed Generation Placement in Power Distribution Networks: Models, Methods, and Future Research. *IEEE Trans. Power Syst.* **2013**, *28*, 3420–3428. [CrossRef]
10. Georgilakis, P.; Hatziargyriou, N.D. A review of power distribution planning in the modern power systems era: Models, methods and future research. *Electr. Power Syst. Res.* **2015**, *121*, 89–100. [CrossRef]
11. Li, R.; Wang, W.; Chen, Z.; Jiang, J.; Zhang, W. A Review of Optimal Planning Active Distribution System: Models, Methods, and Future Researches. *Energies* **2017**, *10*, 1715. [CrossRef]
12. Kalambe, S.; Agnihotri, G. Loss minimization techniques used in distribution network: Bibliographical survey. *Renew. Sustain. Energy Rev.* **2014**, *29*, 184–200. [CrossRef]
13. Sultana, U.; Khairuddin, A.B.; Aman, M.M.; Mokhtar, A.; Zareen, N. A review of optimum DG placement based on minimization of power losses and voltage stability enhancement of distribution system. *Renew. Sustain. Energy Rev.* **2016**, *63*, 363–378. [CrossRef]
14. Sirjani, R.; Jordehi, A.R. Optimal placement and sizing of distribution static compensator (D-STATCOM) in electric distribution networks: A review. *Renew. Sustain. Energy Rev.* **2017**, *77*, 688–694. [CrossRef]
15. Kazmi, S.A.A.; Shahzad, M.K.; Shin, D.R. Multi-Objective Planning Techniques in Distribution Networks: A Composite Review. *Energies* **2017**, *10*, 208. [CrossRef]
16. Kazmi, S.A.A.; Shahzad, M.K.; Khan, A.Z.; Shin, D.R. Smart Distribution Networks: A Review of Modern Distribution Concepts from a Planning Perspective. *Energies* **2017**, *10*, 501. [CrossRef]
17. Chen, T.-H.; Huang, W.-T.; Gu, J.-C.; Pu, G.-C.; Hsu, Y.-F.; Guo, T.-Y. Feasibility Study of Upgrading Primary Feeders From Radial and Open-Loop to Normally Closed-Loop Arrangement. *IEEE Trans. Power Syst.* **2004**, *19*, 1308–1316. [CrossRef]
18. Kumar, P.; Gupta, N.; Niazi, K.R.; Swarnkar, A. A Circuit Theory-Based Loss Allocation Method for Active Distribution Systems. *IEEE Trans. Smart Grid* **2017**, *10*, 1005–1012. [CrossRef]
19. Kazmi, S.A.A.; Shahzaad, M.K.; Shin, D.R. Voltage Stability Index for Distribution Network connected in Loop Configuration. *IETE J. Res.* **2017**, *63*, 1–13. [CrossRef]
20. Kazmi, S.A.A.; Shin, D.R. DG Placement in Loop Distribution Network with New Voltage Stability Index and Loss Minimization Condition Based Planning Approach under Load Growth. *Energies* **2017**, *10*, 1203. [CrossRef]
21. Buayai, K.; Ongsakul, W.; Nadarajah, M. Multi?objective micro?grid planning by NSGA?II in primary distribution system. *Eur. Trans. Electr. Power* **2011**, *22*, 170–187. [CrossRef]
22. Cortes, C.A.; Contreras, S.F.; Shahidehpour, M. Microgrid Topology Planning for Enhancing the Reliability of Active Distribution Networks. *IEEE Trans. Smart Grid* **2017**, *9*, 6369–6377. [CrossRef]
23. Che, L.; Zhang, X.; Shahidehpour, M.; AlAbdulwahab, A.; Al-Turki, Y. Optimal Planning of Loop-Based Microgrid Topology. *IEEE Trans. Smart Grid* **2016**, *8*, 1771–1781. [CrossRef]

24. Sharma, A.K.; Murty, V.V.S.N. Analysis of Mesh Distribution Systems Considering Load Models and Load Growth Impact with Loops on System Performance. *J. Inst. Eng. Ser. B* **2014**, *95*, 295–318. [CrossRef]
25. Murty, V.; Kumar, A. Optimal placement of DG in radial distribution systems based on new voltage stability index under load growth. *Int. J. Electr. Power Energy Syst.* **2015**, *69*, 246–256. [CrossRef]
26. Wang, W.; Jazebi, S.; De Leon, F.; Li, Z. Looping Radial Distribution Systems Using Superconducting Fault Current Limiters: Feasibility and Economic Analysis. *IEEE Trans. Power Syst.* **2017**, *33*, 2486–2495. [CrossRef]
27. Chen, T.-H.; Lin, E.-H.; Yang, N.-C.; Hsieh, T.-Y. Multi-objective optimization for upgrading primary feeders with distributed generators from normally closed loop to mesh arrangement. *Int. J. Electr. Power Energy Syst.* **2013**, *45*, 413–419. [CrossRef]
28. Alvarez-Herault, M.-C.; N'Doye, N.; Gandioli, C.; Hadjsaid, N.; Tixador, P. Meshed distribution network vs reinforcement to increase the distributed generation connection. *Sustain. Energy, Grids Networks* **2015**, *1*, 20–27. [CrossRef]
29. Gupta, A.; Kumar, A. Optimal placement of D-STATCOM using sensitivity approaches in mesh distribution system with time variant load models under load growth. *Ain Shams Eng. J.* **2018**, *9*, 783–799. [CrossRef]
30. Kazmi, S.A.A.; Janjua, A.K.; Shin, D.R. Enhanced Voltage Stability Assessment Index Based Planning Approach for Mesh Distribution Systems. *Energies* **2018**, *11*, 1213. [CrossRef]
31. Kazmi, S.A.A.; Ahmad, H.W.; Shin, D.R. A New Improved Voltage Stability Assessment Index-centered Integrated Planning Approach for Multiple Asset Placement in Mesh Distribution Systems. *Energies* **2019**, *12*, 3163. [CrossRef]
32. Taher, S.A.; Afsari, S.A. Optimal location and sizing of DSTATCOM in distribution systems by immune algorithm. *Int. J. Electr. Power Energy Syst.* **2014**, *60*, 34–44. [CrossRef]
33. Devi, S.; Geethanjali, M. Optimal location and sizing determination of Distributed Generation and DSTATCOM using Particle Swarm Optimization algorithm. *Int. J. Electr. Power Energy Syst.* **2014**, *62*, 562–570. [CrossRef]
34. Tolabi, H.B.; Ali, M.H.; Rizwan, M. Simultaneous Reconfiguration, Optimal Placement of DSTATCOM, and photovoltaic Array in Distribution System Based on Fuzzy-ACO Approach. *IEEE Trans. Sustain. Energy* **2015**, *6*, 210–218. [CrossRef]
35. Devabalaji, K.R.; Ravi, K. Optimal size and sitting of multiple DG and DSTATCOM in radial distribution system using Bacterial Foraging Optimization Algorithm. *Ain Shams Eng. J.* **2016**, *7*, 959–971. [CrossRef]
36. Murty, V.V.S.N.; Kumar, A. Impact of D-STATCOM in distribution systems with load growth on stability margin enhancement and energy savings using PSO and GAMS. *Int. Trans. Electr. Energy Syst.* **2018**, *28*, e2624. [CrossRef]
37. Yuvaraj, T.; Ravi, K. Multi-objective simulations DG and DSTATCOM allocation in radial distribution networks using cuckoo searching algorithm. *Alex. Eng. J.* **2018**, *57*, 2729–2742. [CrossRef]
38. Kamble, S.G.; Vadirajacharya, K.; Patil, U.V. Decision Making in Power Distribution System Reconfiguration by Blended Biased and Unbiased Weightage Method. *J. Sens. Actuator Networks* **2019**, *8*, 20. [CrossRef]
39. Paterakis, N.G.; Mazza, A.; Santos, S.; Erdinc, O.; Chicco, G.; Bakirtzis, A.; Catalao, J.P.S. Multi-objective reconfiguration of radial distribution systems using reliability indices. *IEEE Trans. Power Syst.* **2016**, *31*, 1048–1062. [CrossRef]
40. Kamble, S.G.; Vadirajacharya, K.; Patil, U.V. Comparison of Multiple Attribute Decision-Making Methods—TOPSIS and PROMETHEE for Distribution Systems. In *Computing, Communication and Signal Processing*; Springer Science and Business Media LLC: Berlin/Heidelberg, Germany, 2019; pp. 669–680.
41. Kamble, S.; Patil, U. Performance Improvement of Distribution System by Using PROMETHEE—Multiple Attribute Decision Making Method. *Adv. Intell. Syst. Res.* **2017**, *137*, 493–498.
42. Sattarpour, T.; Nazarpour, D.; Golshannavaz, S.; Siano, P. A multi-objective hybrid GA and TOPSIS approach for sizing and siting of DG and RTU in smart distribution grids. *J. Ambient. Intell. Humaniz. Comput.* **2016**, *9*, 105–122. [CrossRef]
43. Mazza, A.; Chicco, G.; Russo, A. Optimal multi-objective distribution system reconfiguration with multi criteria decision making-based solution ranking and enhanced genetic operators. *Int. J. Electr. Power Energy Syst.* **2014**, *54*, 255–267. [CrossRef]
44. Tanwar, S.S.; Khatod, D.K. Techno-economic and environmental approach for optimal placement and sizing of renewable DGs in distribution system. *Energy* **2017**, *127*, 52–67. [CrossRef]
45. Sultana, S.; Roy, P. Multi-objective quasi-oppositional teaching learning based optimization for optimal location of distributed generator in radial distribution systems. *Int. J. Electr. Power Energy Syst.* **2014**, *63*, 534–545. [CrossRef]

46. Bayat, A.; Bagheri, A. Optimal active and reactive power allocation in distribution networks using a novel heuristic approach. *Appl. Energy* **2019**, *233*, 71–85. [CrossRef]
47. Kumar, S.; Mandal, K.K.; Chakraborty, N. Optimal DG placement by multi-objective opposition based chaotic differential evolution for techno-economic analysis. *Appl. Soft Comput.* **2019**, *78*, 70–83. [CrossRef]
48. Kazmi, S.A.A.; Hasan, S.F.; Shin, D.-R. Multi criteria decision analysis for optimum DG placement in smart grids. In Proceedings of the 2015 IEEE Innovative Smart Grid Technologies—Asia (ISGT ASIA), Bangkok, Thailand, 3–6 November 2015; pp. 1–5.
49. Vita, V. Development of a Decision-Making Algorithm for the Optimum Size and Placement of Distributed Generation Units in Distribution Networks. *Energies* **2017**, *10*, 1433. [CrossRef]
50. Espie, P.; Ault, G.; Burt, G.M.; McDonald, J. Multiple criteria decision making techniques applied to electricity distribution system planning. *IEE Proc. Gener. Transm. Distrib.* **2003**, *150*, 527. [CrossRef]
51. Arshad, M.A.; Ahmad, S.; Afzal, M.J.; Kazmi, S.A.A. Scenario Based Performance Evaluation of Loop Configured Microgrid Under Load Growth Using Multi-Criteria Decision Analysis. In Proceedings of the 14th International Conference on Emerging Technologies (ICET), Islamabad, Pakistan, 21–22 November 2018; pp. 1–6.
52. Javaid, B.; Arshad, M.A.; Ahmad, S.; Kazmi, S.A.A. Comparison of Different Multi Criteria Decision Analysis Techniques for Performance Evaluation of Loop Configured Micro Grid. In Proceedings of the 2019 2nd International Conference on Computing, Mathematics and Engineering Technologies (iCoMET), Sukkur, Pakistan, 30–31 January 2019; pp. 1–7.
53. Quadri, I.A.; Bhowmick, S.; Joshi, D. Multi-objective approach to maximize loadability of distribution networks by simultaneous reconfiguration and allocation of distributed energy resources. *IET Gener. Transm. Distrib.* **2018**, *12*, 5700–5712. [CrossRef]
54. Kansal, S.; Kumar, V.; Tyagi, B. Hybrid approach for optimal placement of multiple DGs of multiple types in distribution networks. *Int. J. Electr. Power Energy Syst.* **2016**, *75*, 226–235. [CrossRef]
55. Sultana, S.; Roy, P. Krill herd algorithm for optimal location of distributed generator in radial distribution system. *Appl. Soft Comput.* **2016**, *40*, 391–404. [CrossRef]
56. Muthukumar, K.; Jayalalitha, S. Optimal placement and sizing of distributed generators and shunt capacitors for power loss minimization in radial distribution networks using hybrid heuristic search optimization technique. *Int. J. Electr. Power Energy Syst.* **2016**, *78*, 299–319. [CrossRef]
57. Kashyap, M.; Kansal, S.; Singh, B.P. Optimal installation of multiple type DGs considering constant, ZIP load and load growth. *Int. J. Ambient. Energy* **2018**, 1–9. [CrossRef]

MDPI
St. Alban-Anlage 66
4052 Basel
Switzerland
Tel. +41 61 683 77 34
Fax +41 61 302 89 18
www.mdpi.com

Energies Editorial Office
E-mail: energies@mdpi.com
www.mdpi.com/journal/energies

www.ingramcontent.com/pod-product-compliance
Lightning Source LLC
LaVergne TN
LVHW070904170726
843515LV00005B/703

* 9 7 8 3 0 3 9 3 6 3 6 6 7 *